THE TECHNOLOGICAL SPECIALIZATION OF ADVANCED COUNTRIES

The Technological Specialization of Advanced Countries

A Report to the EEC on International Science and Technology Activities

Preface by Jacques Delors,
President of the Commission of the European Communities

DANIELE ARCHIBUGI
Istituto di Studi sulla Ricerca e Documentazione Scientifica (ISRDS),
National Research Council, Rome

and

MARIO PIANTA
Istituto di Studi sulla Ricerca e Documentazione Scientifica (ISRDS),
National Research Council, Rome

Springer Science+Business Media, B.V.

Library of Congress Cataloging-in-Publication Data

Archibugi, Daniele.
The technological specialization of advanced countries : a report to the EEC on international science and technology activities . Daniele Archibugi, Mario Pianta ; preface by Jacques Delors.
p. cm.
Includes bibliographical references and index.
ISBN 0-7923-1750-5 (alk. paper)
1. Technology--European Economic Community countries. 2. Science--European Economic Community countries. I. Pianta, Mario, 1956- . II. European Economic Community. III. Title.
T26.A1A73 1992 92-8629

ISBN 978-94-015-8001-4 ISBN 978-94-015-7999-5 (eBook)
DOI 10.1007/978-94-015-7999-5

Printed on acid-free paper

Table of Contents

List of Tables

List of Figures

List of Abbreviations

CEC:	Commission of the European Communities
CNR:	Consiglio Nazionale delle Ricerche, Italy
EEC:	European Economic Community
EPO:	European Patent Office
GDP:	Gross Domestic Product
GERD:	Gross Domestic Expenditure on Research and Development
IPC:	International Patent Classes
ISRDS:	Istituto di Studi sulla Ricerca e Documentazione Scientifica, Italy
MSTI:	Main Science and Technology Indicators of the OECD
NACE:	General Industrial Classification of Economic Activities
OECD:	Organization for Economic Cooperation and Development
R&D:	Research and Development
SCI:	Science Citation Index
SIC:	Standard Industrial Classification
SITC:	Standard International Trade Classification
SPRU:	Science Policy Research Unit of the University of Sussex
SRCA:	Index of Scientific Revealed Comparative Advantage
S&T:	Science and Technology
TRCA:	Index of Technology Revealed Comparative Advantage
WIPO:	World Intellectual Property Organization

Preface

Science and technology are essential for Europe's future. They are essential not only for its economic prosperity and industrial competitiveness but also for the quality of life, the environment and the social and cultural climate.

With the adoption of the Single European Act, the importance of science and technology and the need for Community involvement were formally recognised and their ever increasing importance will mean that they are certain to figure larger in any future revisions of the Treaties.

Although Community spending on science and technology is small compared with what the Member States themselves spend, the actual sums are significant and have been growing steadily over the last few years. By observing the principle of subsidiarity to Member State activities, Community R&D action gives rise to considerable added value at the European level. Effective subsidiarity requires that priorities be carefully chosen and that the effects of Community activities be measured in order to guide future action.

We need a thorough understanding of the scientific and technical environment in Europe, its evolution and the strength and weaknesses of Europe compared with its competitors. In addition, a knowledge of how science and technology influences industrial, social and economic development is essential to guarantee the effectiveness of Community action. The Monitor programme was launched in 1989 precisely to gain such understanding and to obtain such knowledge.

The development of appropriate indicators is one of the first steps needed in the management of science and technology and the results presented in this report are from one of the first studies funded by the Monitor programme, jointly with the Italian National Research Council. It is part of a series of actions carried out by networks of European experts with the aim of developing R&D evaluation methodologies. In addition, the work aims to define quantitative indicators and to assess the impact of science and technology on industrial competitiveness and regional development.

The work of Archibugi and Pianta reported here uses indicators such as R&D expenditure, patents and scientific publications to explore the characteristics of science and technology in Europe and in other advanced countries. From the critical study of the results, the areas of relative advantage in individual countries and in the Community as a whole clearly emerge as do changes with time and over different sectors.

Overall, this report provides vital information on national performance in innovation. It will be of particular importance for defining measures to advance the Community's scientific and technological position in the world and to evaluate the results of policies already undertaken. I am thus happy that the Commission has sponsored this study and its publication and hope that it may be of interest and use to a wide audience.

JACQUES DELORS,
President of the Commission of the European Communities

Acknowledgements

This book on the science and technology activities of advanced countries and their sectoral strengths and weaknesses is based on a research project carried out over the years 1989–91 at the Institute for Studies on Scientific Research and Documentation of the Italian National Research Council in Rome. The research was financed by the Commission of the European Communities, Directorate-General XII, Science, Research and Development, Service Evaluation (*Development of indicators to measure European R&D potential, impact, regional imbalances and need for co-operation*, contract No. EVAL-0051-I (CH)), and by the Italian National Research Council, Project on Technology Transfer.

This research report was jointly written by Daniele Archibugi, research director, and by Mario Pianta, researcher at the ISRDS-CNR. Rinaldo Evangelista and Roberto Simonetti were part of the project team and provided useful research assistance and statistical elaborations. Patrizia Principessa was in charge of the data bases and of their elaboration. Maurizio Vichi gave important suggestions on the statistical analysis. Paolo Capasso contributed to the preparation of figures. A valuable contribution was provided by Giorgio Sirilli, and useful comments came from Sergio Cesaratto and Alberto Silvani, all colleagues of ours at the ISRDS-CNR. We are indebted to all of them for their contribution. A number of preliminary reports with analysis and elaborations on specific data were produced in the course of this project. They will be referred to as sources of the data summarized here, and are available on request[1].

During this research, we benefited from discussion with several collegues and experts. We wish to thank Luigi Massimo, Head of the Service Research Evaluation of the D.G. XII of the Commission of the European Communities for his suggestions and contributions. Clara De La Torre and Grant Lewison of D.G. XII provided the research team with important feed-back. Collegues from the SPEAR Network on S&T indicators for Evaluation also provided valuable comments upon earlier drafts. We especially wish to thank Hariolf Grupp, Kim Møller, and Antony van Raan for detailed criticism and for their participation at the meeting held in our Institute in January 1990. A work in progress of this report was also presented and discussed at the Commission headquarters in Brussels in April 1990.

We thank our ISRDS-CNR colleagues for their collaboration in this project and the director, Prof. Paolo Bisogno. The friendly environment of the Institute was of great help when we were particularly overloaded. The ISRDS-CNR also hosted two seminars on the patterns of technological specialization, and we benefited from the discussions among the participants.

Keith Pavitt and Pari Patel of the Science Policy Research Unit of the University of Sussex and John Cantwell of the Department of Economics of the University of Reading were part of our "invisible college" on the use of patenting as a technological indicator. CHI Research provided the data on patent counts and citations in the US and on paper counts and paper citations, which are extensively employed in this report. We also appreciated the aid of the OECD, Division Science, Technology, and Industry Indicators, in providing updated S&T data. The collaboration of the US National Science Foundation and of other international institutions is also appreciated.

Preliminary findings of the research have been presented and discussed in various seminars and conferences held in the following institutions: Department of Economics of the University of Reading; National Science Foundation in Washington D.C.; Institute for Economics and Planning of the Roskilde University Centre; Institute for International Economics and Management of the Copenhagen Business School; Department of Applied Economics of the University of Cambridge; Conference of the International Studies Association held in London; Conference on European Competitiveness held at WZB in Berlin; Meeting of the Schumpeter Society held in Virignia; Conferences of the European Association for Research in Industrial Economics held in Lisbon and Ferrara; Meeting of the European Association for Evolutionary Political Economy held in Florence. Articles with various works in progress of this research have appeared in *Research Policy*, *Scientometrics* and *L'Industria*. Detailed comments were provided, at different stages, by Nicola Acocella, Giovanni Amendola, David Audretsch, Massimo Di Matteo, Chris Freeman, Paolo Guerrieri, Kirsty Hughes, Alfred Kleinknecht, Antonio Perrucci and anonymous referees. We are very grateful for the comments received; obviously, the responsibility for errors and judgements remains our own.

Note

1. They include: D. Archibugi and M. Pianta, 1989, *The Technological Specialization of Advanced Countries*, Ad Interim Report, Commission of European Communities, September 1989; M. Pianta et al., 1990, *La Specializzazione Scientifica dei Paesi piu' Avanzati: Un'Analisi degli Indicatori Bibliometrici*, ISRDS-CNR, September 1990; R. Simonetti, *Un'Analisi per Settori della Specializzazione Tecnologica: I Dati sui Brevetti dei Paesi Avanzati*, ISRDS-CNR, September 1990.

CHAPTER 1

Introduction and Summary

This book investigates the structure and the changes over time of the technological specialization of advanced countries. Innovation is generally regarded as one of the key determinants of national economic performance and competitiveness. During the 1980s much attention was devoted to understanding its dynamics and impact. An increasing amount of studies have been devoted to the aggregate quantity of resources devoted to Science and Technology (S&T). However, less attention has been paid to a systematic analysis of national activities at the sectoral level. One stream of studies has pointed out that technological change is highly diversified across sectors and countries,[1] suggesting that the importance of the innovations of each country differ strongly according to their sectoral specialization. We shall here explore the relationship between general tendencies in the S&T system on the one hand and national patterns of sectoral specialization on the other.

The same subject has already been considered, from a different perspective, by scholars in the field of international economics and management. It is often pointed out, for example, that the process of globalization in the economy is parallel to increasing sectoral specialization in production (for a recent overview, see Porter (1990)). In other words, the international integration of economic systems leads to the emergence of specific fields of national specialization in industrial production. The process of "specialization" is the key issue of this report. By scientific or technological specialization we mean the structure of the distribution of activities across different sectors of a country. Besides the obvious differences in the aggregate amount of resources devoted to S&T by individual nations, it is important to identify each country's sectors of excellence in order: i) to describe its relative specialization, ii) to assess the quality of the aggregate performance and iii) to identify the likely direction of future developments. The study of specialization provides valuable information on the role of, and tendencies in, the S&T system. Since advanced countries are upgrading the technological level of their economies, it becomes crucial to assess whether they are concentrating efforts in the same fields (either in research or innovation) or focusing in different selected areas where economies of scale and scope can be exploited.

For an area such as the EEC, characterized by many small and medium-sized countries and by a process of economic integration, these issues are particularly important. Given the small size of European countries compared to the United

States and Japan, the process of economic and technological specialization in Europe has assumed a particular role in shaping changes in industrial structure, national performance and industrial strategies. The prospect of greater economic integration marked by the development of the single European market by 1993 opens up new possibilities for achieving a critical mass of S&T resources in many areas. Both the process of specialization and its relation to the scale of technological capacity are important for the present conditions of the EEC countries. In fact, national specialization patterns in S&T are related to qualitative interactions among countries. The more specialized a country is in selected areas, the greater the exchange and cooperation needed with other countries in order to take advantage of national sectors of strength and to gain access to advanced know how in areas of national weakness.

Technological specialization and international integration are processes which are strongly affected by the activity of firms operating on a global scale.[2] Multinational firms are among the most important producers of techology in the advanced countries, and their strategies are of great importance in shaping the patterns of specialization. Throughout this report, however, countries will be the unit of analysis, and the specialization considered will refer to all activities performed by organizations located within the national borders. The particular role and behaviour of foreign-owned firms and of affiliates in other countries is not considered here.[3]

The use of national data is, however, of relevance. Firstly, it is widely accepted that "national systems of innovation" have emerged, with individual countries characterized by a specific pattern of technological accumulation, industrial structure, and by a peculiar set of institutions supporting and regulating technological change. Secondly, while the economic operations of firms are highly mobile across borders, many factors contributing to innovative performance, including labour and the role played by the scientific communities, are much more country-specific and can be measured by national indicators. Furthermore, it is at the national level that major technology policy decisions are made, creating the institutional framework which allows the development of invention and innovation.

A number of key issues and processes have been pointed out in recent years in economic as well as science and technology studies, and will be discussed in Chapter 2. We shall also be considering the importance of national systems of innovation and the increasing international integration of science and technology, which have led to new roles for national institutions, to growing cooperation among national scientific communities, to government-sponsored international research programmes and to increasing inter-firm technical agreements. A few questions on these patterns and on company strategies will be addressed, providing the background for the empirical analysis of the following Chapters.

Chapter 3 provides an overview of the S&T activities carried out in the OECD countries. An analysis of aggregate national efforts is required to put the study of sectoral specialization into context. In fact, the emergence of relative strengths or weaknesses has to be related to the characteristics and size of national S&T activities. The value and limitations of the main indicators used in this report is also discussed. They include: i) the resources devoted to R&D, ii) patenting, and iii) bibliometric indicators. These indicators describe the S&T intensity of national economies and are generally strongly and positively correlated at the aggregate level. However, the differences among countries highlighted by S&T indicators reflect the nature of national innovation systems. In some countries national priority is directed towards "scientific" effort, while other national systems give priority to technological activities more directly linked to market activities. These differences are discussed in Chapter 3. In the same Chapter the tendencies over time in S&T activities are also examined. R&D inputs have grown considerably in the OECD countries, although signs of a slow-down have been seen in recent years. In the majority of countries a growing propensity to appropriate the benefits of their innovations on the global scale has also emerged, as is indicated by the fast growth of patents obtained in foreign markets.

Chapters 4 and 5 present a detailed description of national sectoral strengths and weaknesses in technological activity. This mapping is carried out by means of several patent-based indicators. Chapter 4 examines patents registered during the 1980s in four patent offices: the US, West Germany, France, and the European Patent Office; data are disaggregated in terms of the two digit International Patent Classes. This Chapter considers the patenting activity of eleven countries: the United States, Japan, West Germany, France, the United Kingdom, Italy, the Netherlands, Belgium, Switzerland, Sweden and Canada.

We have focused on various patent institutions for several reasons. While the US patent system has often been taken as a basis for international comparisons at the sectoral level, until recently much less attention has been paid to the consistency of the national profiles of specialization[4] resulting from data drawn from different patent institutions. In Chapter 4 the stability of national patterns will be tested, showing that a significant difference emerges between the specialization profiles based on patents registered in the domestic and in the external markets. This is not surprising, since each patent institution is characterized by a large number of domestic inventions, many of which are of low quality and impact, and seldom become actual innovations. The effect that patent behaviour in the internal market has on a country's specialization profile can be defined as the "domestic market effect" in patenting activity.

Chapter 5 presents further analysis of technological specialization, considering a larger number of countries than in Chapter 4 (all the EEC member countries, the US, Japan, Canada, Switzerland and Sweden) over the period 1975–88 and according to a classification (the Standard Industrial Classes) which is more closely related to economic indicators. In the same Chapter, an indicator of the impact of patents is also considered, i.e. patent citations. Since the US patent system was shown in Chapter 4 to provide a reliable picture of the profiles of technological specialization for most countries (the US excluded), this Chapter focuses on patents granted in the US only.

The results of Chapter 5 show the cross-country differences in the average impact of patented inventions. Japanese patents seem to have a higher impact than those of any other country, the US included. European patents, on the contrary, are cited less often than the world average. In the same Chapter it is also shown that the sectoral specialization of the different countries, measured on both patent counts and patent citations, is highly correlated; not surprisingly, for the majority of countries the two indicators highlight the same sectoral strengths and weaknesses. Strong positive correlation coefficients are also found over time, providing additional support for the hypothesis of the cumulative nature of national technological capabilities (see Pavitt (1987), Cantwell (1989)).

In Chapter 6 a further analysis of countries' technological specialization is carried out considering the rates of change registered in 118 patent sub-classes. Fast growing patent classes are likely to be associated with original scientific and technological developments or to increasing competition among firms for the exploitation of specific innovations. Possession of an above average number of patents in fast growing classes represents an advantage for a country's innovative activity. National sectoral specialization is then related to the innovative dynamism in the fields of national strength, as indicated by the rates of change of total patents in each sub-class. This indicator provides information on the quality of each country's technological specialization. The results of Chapter 6 show that the sectoral strengths of European countries are often found in declining rather than in fast growing classes. On the contrary, Japan's share of patents in fast growing classes is higher than its share of total patents.

Chapter 7 assesses sectoral specialization in science by means of bibliometric indicators. The activities of national scientific communities provide one of the sources of innovation, and the relationship between specialization in scientific and technological activities is tentatively explored. Obviously, not all the scientific fields are equally related to technological or economic processes. The scientific communities of some countries have a greater propensity to focus on fields which are more directly linked to economic performance, indicating the prominent role played by scientific institutions in shaping the patterns of specialization in technology. This is the case in Japan, where a large proportion of the efforts of the national scientific community is channelled into scientific fields closely related to economic or technological activities.

In addition to the descriptive mapping of each country's sectoral strengths and weaknesses, more general analysis is required on the process of specialization. Some countries concentrate their efforts in selected areas, while others tend to distribute their S&T resources more uniformly. These issues are considered in Chapter 8, which contains the analytical core of this report. The issue of defining the degree of specialization of individual countries is discussed and an appropriate methodology for empirical analysis is developed. Strong differences in the degree of specialization across countries are found both in scientific and in technological activities. While several factors contribute to these variations, the amount of national resources devoted to S&T emerges as the most important. In fact, a clear-cut inverse relationship emerges between national technological size (measured by R&D expenditure) and the degree of specialization in technology (measured as the dispersion across sectors in patent based indicators). Only large countries can afford to spread their activities across most technological fields, while small and medium sized countries are to some extent forced to specialize in more narrow niches.

The same inverse relationship also emerges between the size of each country's scientific activities (measured by the number of researchers of the non-business sector) and their degree of specialization (measured as the dispersion across fields shown by bibliometric indicators). This inverse relationship seems, however, to be less marked in scientific than in technological activities. This result reveals a basic difference between science and technology: the non proprietary nature of the former makes it possible for countries of all sizes to distribute their activities more evenly.

In some countries the degree of specialization was found to be substantially higher than one would expect from the size of their S&T activities. These national differences are equally reflected in both technology and science indicators. Japan and, to a lesser extent, Italy present a higher than average degree of specialization in scientific and technological activities. Conversely, Britain and France have a comparatively low level of specialization. These patterns can be seen as the result of substantial differences in terms of scientific and technological accumulation and of the diverging strategies followed by firms and governments. Trends in the degree of specialization in both science and technology have also been explored and a clear tendency towards increasing specialization in technology is evident in most countries. This may be regarded as the other side of the globalization process: as countries increase their international integration, they are forced to concentrate their effors in the sectors where excellence can be achieved.

An opposite trend has occurred in the degree of specialization shown by scientific activities, which has in fact decreased in the majority of countries. These diverging tendencies shown by scientific and technological indicators raise the problem of the different nature of these two activities. The non proprietary nature of scientific results has allowed the majority of countries to expand their activities in the sectors of their relative weakness. On the contrary, the nature of technological change makes it easier, and more advantageous, for countries to concentrate on their sectoral advantages.

The analysis of the degree of specialization in technology is further explored, at the sectoral level, in Chapter 9. Specialization has increased in most sectors and the clusters related to specialized equipment and machinery constitute the bulk of this group. The same high and growing specialization is found in production-intensive sectors, thus reflecting the uneven distribution of productive capacities across countries. The technological classes which show the opposite pattern, i.e. a low and decreasing degree of specialization, are on the contrary to be found in some electronic fields and in other very pervasive sectors.

Chapter 10 explores similarities and differences of national profiles of innovative activities and the impact specialization has on national performances. First, a measure of the distance between pairs of countries is developed; the technological profiles of larger countries appear rather similar, as activities are spread more uniformly across fields. Smaller countries concentrate their efforts in fewer sectors, showing greater distance from larger ones. More importantly, their sectors of strength differ from country to country, resulting in very high technological distances. While size represents a constraint on the number of fields in which a country can be active, it rarely prevents the development of a national specialization in a particular sector.

Second, the analysis of the impact of specialization on national performances shows that a faster growth of patenting activities and industrial production is associated to higher degrees of technological specialization (weighted to account for country's size). While many other factors, including the size and nature of technological activities, play a role in explaining national performances, the results suggest that there is a specific advantage for countries to concentrate their efforts in selected technological fields, regardless of the type of sectors involved. Even a specialization in traditional low technology fields appears better than no specialization at all.

Finally, the conclusions in Chapter 11 bring together the different findings and provide an overview of the major patterns in the science and technology specialization of advanced countries. Some policy implications are discussed in the light of the patterns of increasing specialization and international integration of technological activities. For individual countries there is a need to balance the advantages of maintaining already established areas of technological strength on the one hand, and, on the other hand, the long term necessity to develop activities and specializations in new fields, where uncertainty, competition, but also potential payoffs, are greater. An appropriate targeting of sectoral technology policies and further efforts for international cooperation can improve the contribution of technological activities to economic growth.

Notes

1. The paradigmatic works of this new stream in the field of the economics of technological change include Rosenberg (1976, 1982), Nelson and Winter (1982), Freeman (1982), Pavitt (1984, 1987), Dosi et al. (1988).
2. On the role of multinational firms in the global economy, see, among others, Dunning (1988), Holland (1987), Van Tulder and Junne (1988). For a recent survey, see the OECD report of the Technology and Economy Program, OECD (1990).
3. The research teams of the Science Policy Research Unit of the University of Sussex and of the Department of Economics of the University of Reading have investigated the role of multinational firms in the national patterns of technological specialization. See Patel and Pavitt (1989a), and Cantwell and Hodson (1990).
4. Throughout this report, by "profile of specialization" is meant the vector showing the relative distribution of activities of a country in all the sectors according to one of the databases considered.

CHAPTER 2

Structural Change and International Strategies in Science and Technology

2.1. The Nature and Role of Technological Change

The development and diffusion of new technologies is a major aspect of the current transformations in advanced countries. Innovation in products, processes and forms of organization is a critical factor for greater productivity, competitiveness, growth and employment, as indicated by a large number of theoretical and empirical analyses.[1]

Mastering technological innovation is, however, not an easy task. National success in innovative activity requires the combination of different factors, ranging from a good research base to institutions supporting technical advance, from adequate managerial skills to rapid learning ability. Technological change is no longer seen as a linear process, from scientific research, to technological development, to the engineering of new products and processes. The mechanisms of innovation are much more complex. While an active and widespread scientific base is crucial for the development of innovations, they tend to emerge from an intricate web of relationships, building on the existing technological accumulation, with firm- and country-specific patterns.

Institutions supporting and directing technical change are a key element in a country's technology policy, and an important role is also played by the particular social and cultural environment in which innovation policies are implemented. In turn, the way new technologies are developed and introduced deeply affects firms, labour, the industry structure, and society as a whole.[2]

To keep abreast of international competition, each country has to increase its effort to produce new technology and to adapt and disseminate innovations. In an increasingly integrated world economy, desire to reap the benefits of technological change is a key element shaping company strategies and government policies. The factors required to innovate successfully are, however, unevenly distributed across countries, creating substantial differences in both the quantity and the nature of the innovations produced. Nations where barriers against innovation are higher will lose ground to their competitors.

A major development of the last few decades is that technological change is taking place increasingly at the international level and its impact on economic and social life is also perceived globally. While science and technology are certainly not the only activities to have increased their international dimension (see Holland (1987)), they are one of the spheres of economic life where the process of globalisation has been more intense and explicit.

The changes in the nature of the science and technology systems, the role of national government policies, and the international strategies that firms and governments are developing will be reviewed in this Chapter so as to provide a preliminary description of the processes of specialization to be investigated in the following Chapters. It presents an overview of the main themes to which the evidence provided in the rest of the book has to be related.

2.2. The New Strategy of Firms and the Internationalization of Technology

Firms are among the main producers and users of technological knowledge and innovation. In the last two decades, the restructuring of industrial production in advanced countries has been combined with the development of new technologies, thus leading to a system which is more capital and skill intensive and labour saving, but also more flexible and integrated at the world scale. Innovations have been developed according to specific criteria of cost, performance and quality, which greatly vary across firms, industries and countries.

Industrial restructuring at a global level has led to a growing international orientation of large firms. Intra-firm trade with foreign subsidiaries has increased, together with their share of total national imports and exports.[3] Firms have also expanded foreign direct investment and technology trasfer.[4]

The importance of multinational firms in the technological activities of advanced countries has already been documented. According to the OECD, multinational firms account for 75% of all industrial R&D in OECD countries. The combined R&D expenditure of the five largest firms of the US, Japan and the UK is greater than the total resources, both private and public, devoted to R&D in the 15 OECD countries with the lowest level of R&D activities (OECD (1985a, p. 63). See also Cantwell (1989), Patel and Pavitt (1989a)).

Besides the size of the largest firms' innovative activities, the international scale of their research should be noted. A study by Mansfield et al. (1982, p. 209) found that by the mid–1970s about half of the US firms surveyed had achieved world integration of their R&D activities, and already in the early 1970s US subsidiaries accounted for one-seventh of industrial R&D in the UK and Germany and half in Canada.[5]

However, if we look at the results of innovative processes, the internationalization of technology appears to be left behind by the growth of firms' international production. Patel and Pavitt (1989a) have shown that the world's largest firms continue to produce in their home country the vast majority of innovations, as measured by patents. The technological activity of new affiliates in foreign countries is not the only form of technological internationalization of a firm, and a variety of new cooperative strategies have been developed, as we shall see in section 2.4 below.

These large-firm strategies should be seen in connection with the growing competition in international markets. European and Japanese firms have entered industries where US firms had long enjoyed a critical technological lead. The larger R&D effort for new technologies requires high production volumes, which in turn can offer important economies of scale. The result is increased pressure to export and to enter new markets. These mechanisms contribute to opening up national economies to international trade, which is particularly important in the markets for high technology products. In fact, in most countries the import penetration in R&D-intensive industries has increased very rapidly, and is higher than in other sectors.[6]

The growing internationalization of high technology, and the key role played by firms in this process, does not, however, mean that developments at the national level have become irrelevant. The level and organization of knowledge within each nation is a key factor in the success of firms' innovative activities and important changes can be found in the position of individual countries in the new technological landscape. These aspects are addressed in the following sections.

2.3. National Systems of Innovation and Government S&T Policy

The concept of "national systems of innovation" has been developed in order to account for the different combinations of industrial structure, scientific and technological accumulation, and public policies shaping the innovative activities of advanced countries (see Nelson (1984), Freeman (1987a), Freeman and Lundvall (1988)). These structural differences are often translated into particular "styles" of technological development, linked to specific institutional or market conditions.

The national systems of innovation define the advantages and disadvantages a country offers for the development of innovative activities. Several elements contribute to the characteristics of national systems of innovation. The size of the scientific community, its fields of research and its links with industry play an important role, together with the ability of the educational system to provide highly qualified researchers and technicians. The structure of national industry and the R&D activities of firms are key factors in the success of innovative efforts in specific fields, products and processes. Government policies in all the areas listed above and the specific strategies in R&D and technology are crucial for mobilizing adequate resources and for directing national innovative efforts.

Fresh attention has recently been focused on the country-specific factors behind technological and economic performance. A comparative study on the national innovation systems of advanced countries (Nelson, forthcoming) found that countries do indeed differ in their methods of promoting and regulating technological advance, thus leading to national differences in the patterns of innovative activity and in its economic impact. On similar lines, Porter (1990) stressed the importance of nation-specific elements in shaping company performance.

The activities of the scientific community, firms and government agencies lead to the emergence of a specific set of institutions supporting and regulating technical change and shaping its nature and direction. In some countries, government funded institutions, public laboratories and universities cooperate closely with the business sector; in other countries firms set up their own networks to share know-how and technical information. The amount of resources mobilized, the industrial sectors chosen to become the "national champions" and those targeted for innovation, the importance of the military, the type of institution involved and the criteria for selecting innovations (cost, performance, quality, etc.) are all critical factors in defining the national performance and the technological "style".

An important role is played by government innovation policies, including R&D funding, industrial policies, high technology programmes, control over technology transfer and, more generally, trade policy. The nationally-oriented technology policy of government is, however, in stark contrast to the increasingly international nature of firms' strategies and of technology flows. The impact of government technology policy is likely to be limited to the part of the firm's activity that takes place within the national borders. In general terms, the higher a firm's degree of internationalization, the less power government technology policy is likely to have in affecting its innovative strategy.

On the other hand, the growing international competition makes firms more dependent on government help to develop new technologies and to protect their domestic markets. The result is a set of well-established and evolving relations between firms and government, which take different institutional forms in each country and result in national industrial and trade policies. While government policies set the rules for business activity, they are also strongly influenced by firms' strategies. This is especially so in small countries, where the presence of very few large firms may place serious constraints on national policies.

The result of such patterns is that the internationalization of business strategies does not make nation-specific factors irrelevant. On the contrary, it places new emphasis on countries' endowments, as firms can locate their S&T activities in countries offering particular advantages, e.g. appropriate infrastructures, specific research activity, a more favourable innovative environment. Such behaviour on the part of firms can further strengthen national patterns of technological activity. The way countries have reacted to the process of internationalization and the performance of the firms located within their borders have resulted in changes in the relative positions of individual countries in science and technology activities.

The most marked differences in the national systems of innovations are found among the main advanced regions - Europe, the United States and Japan.[7] A high level of variation can, however, also be found among European countries. The balance between market activities and government policy, the links between science, technology and industry, and the sectors of greatest national activity are key aspects where large differences can be identified also within the twelve EEC countries. The unification of the domestic EEC market after 1992 is likely to change at least some of the entrenched government-firm relations at the national level, and a greater role will be played by Community level policies and institutions supporting technical advance.

Furthermore, closer integration is developing between the EEC and other European countries, some of which, such as Sweden, Switzerland and Austria, have a pattern of S&T activities which is very similar to those of EEC members and very close interaction with them. The former planned economies of East-Central Europe have industrial and innovation systems which are rather different from

Western models, but are involved in rapidly growing cooperation in science and technology with EEC countries. While this changing dimension of Europe creates some uncertainty as to the future outlook for its innovation systems, it also offers new opportunities for its development.

National systems of innovation can be studied from different viewpoints. Freeman (1987a), Freeman and Lundvall (1988) and Nelson (forthcoming) have highlighted the institutional and qualitative aspects which characterize each country. In this book we shall focus on a complementary perspective, i.e. how these qualitative differences are reflected in countries' measurable strengths and weaknesses in S&T.

First, several features of the national innovation system are reflected by the available science and technology indicators which describe the different research intensities of countries and the results achieved by the national systems. Second, the innovation system defines the areas of greatest national effort, and in the associated sectors of science and technology countries tend to develop a relative specialization and an international comparative advantage. Countries where military spending is a high priority show relative strengths in fields such as weapons, aircraft and nuclear research. Countries where an excellence in academic research is matched by strong links with industry may show a relative advantage in science-based technologies. Third, national systems differ in their ability to translate inventions into actual innovations, to spread them rapidly and to turn technological advantages into better economic performance. A high level of R&D effort and high innovative output are necessary but not sufficient conditions for greater competitiveness and productivity. A variety of other characteristics of the innovation system play a role in transforming technological efforts into industrial success.

These national features will be taken into account in interpreting our empirical results. A variety of science and technology indicators, including R&D expenditure, patenting in different institutions and scientific publications, will be used in Chapter 3 to show the convergence between the US, Europe and Japan, a process which is well documented also for indicators of economic and industrial performance.[8] In spite of the different institutional aspects of national systems, technology is becoming a key issue for growth and competitiveness for a larger number of countries.

The phase of European and Japanese "catching-up" with US capabilities has been replaced by the convergence of the major advanced countries at the technological frontier, and a complex pattern of specialization and relative advantages in specific areas of technology is emerging. The analysis of Chapters 4 to 6 will explore the changing sectoral strengths and weaknesses of individual countries in technology, while Chapter 7 will develop a parallel analysis of national activities in science. To a certain extent, the data on countries' S&T activity and their sectoral disaggregation can be interpreted as the quantitative description of the nature and results of each national system of innovation.

A growing degree of specialization is evident in the technological activities of most countries, as Chapters 8 and 10 will show, and countries tend to concentrate their innovative resources in the areas of their greatest advantage. This result highlights the interrelations among individual national systems. Countries differ in the way they organize their technological activities and this is reflected in their sectoral distribution of innovations. These differences increase over time, possibly reinforcing the distinctiveness of national patterns of activity.

Parallel to this, technological knowledge also tends to become more specialized, with new developments and applications in a variety of fields. This leads to a greater concentration of the activities of few countries in most technological sectors, an aspect investigated in Chapter 9. These changes can affect a country's relative position on the international scene. Old specializations may be replaced, institutional aspects can evolve, new upstream or downstream strengths may emerge.[9] From the processes described above, the obvious conclusion is that it has become impossible for a single country to assume technological leadership over the whole range of new technologies. National technology strategies have now to consider the context of convergence at the technogical frontier and need to cope with the growing process of internationalization.

The aim of national strategies cannot be that of achieving (or regaining) overall leadership; nor, for the weaker countries, can it be an indiscriminate effort to imitate the technological model offered by the few leading countries. Rather, the new objective of national technology strategies becomes that of targeting the best position for the national economy in the international division of labour at the technological frontier, and sustaining the relative advantages that have been achieved. A large body of literature (Pavitt (1987), Cantwell (1989)) has in fact shown that successful national innovative performance has to build upon already accumulated capabilities and know-how. This changing context is clearly perceived by the main agents. In fact, it is remarkable that a context of greater convergence and international competition has led to the emergence of more cooperative strategies developed by firms, governments and scientific communities, which are reviewed in the next section.

2.4. The New Cooperative Strategies of Firms, Governments and Scientific Communities

The internationalization of technology and the growing sectoral specialization of the activities of countries and firms have led, over the last decade, to a new pattern of cooperation in innovative activities both across borders and among different institutions, namely research centres, industry and government agencies. Three major aspects of these cooperative strategies will be considered here: international cooperation between firms; the development of high technology programmes combining the efforts of different agents; and the greater international collaboration among scientists.

Cooperation Agreements between Firms

An increasingly important form of internationalization of firms' innovative activities is the development of agreements with foreign companies in the field of new technologies. Several recent studies have shown that such agreements take different forms, including joint development agreements, sharing of research and design resources, investment by large firms as a minority shareholder in small innovative companies.[10] In practical terms, this cooperative strategy can be explained as an attempt to share the higher risks of innovative ventures at the technological frontier, a very different kind of operation from the previous effort to improve already consolidated technologies, moving with incremental advances along a well-defined technological trajectory.

More extensive inter-firm cooperation also allows broader exploitation of the benefits from innovation, as firms try to appropriate the returns from their technology on a global scale. This pressure has increased with the shortening of product cycles, especially in high technology fields, resulting in a more urgent need to bring the new products to larger markets anticipating international competition. Other factors contributing to the new wave of cooperative agreements include the need to define common standards for new technologies and products; the diversity of the possible applications of technological advances in different sectors; the complementarity of developments in different fields; and the "systemic" nature of some innovations (see Van Tulder and Junne (1988)).

Some of these factors leading to greater international inter-firm cooperation are the result of the process of specialization already pointed out. In many fields it is increasingly difficult to find one firm, however large it may be, possessing all the research, development, production and marketing capabilities required for the successful introduction of innovations. Growing complementarities are evident among large and small firms, companies controlling different markets or different segments of a particular technology. The result is an extensive network of inter-firm relations and new patterns of cooperation among firms, where technology and market considerations are combined. The degree of competition is not necessarily reduced, as the object of cooperative arrangements is generally highly focused and represents only a small part of the overall activity of firms.

Not surprisingly, the highest concentration of such agreements can be found in the area of information technologies (see Hagedoorn and Schakenraad (1992a)). The most intense cooperation is between US and Japanese firms on the one hand, and between US and European firms on the other. While intra-European cooperation has increased, trans-Atlantic links still appear to be the priority of European firms in many fields. On the other hand, European technology agreements with Japanese partners are particularly rare.

Inter-firm agreements raise new questions for the analysis of industrial organization. A study on the determinants of strategic technological partnership has indicated that in about one-third of the cases the main motivation is either access to new markets or changes in the market structure. However, in nearly 60% of cases, the firms' main interest is to share R&D costs or to acquire know-how (Hagedoorn and Schakenraad (1992a)). The market determinant can be seen as a step towards concentration or collusion, while the technological determinant may be interpreted as an indicator of a wider and more intensive use of knowledge in products and services.

High Technology Programmes

A special case of cooperative strategies at the national or regional (in the case of Europe) level is that of high technology programmes, which were developed in the 1980s by the US, Japan and the European governments. They have moved beyond the traditional R&D and industrial policies, combining the resources and strategies of governments with the research efforts of scientific communities and the production capacity of firms. They can be defined as large R&D projects targeted to specific technological advances or to the development of industrial or military systems. Major examples include, in chronological order, the Fifth-Generation Computer Programme in Japan (launched in 1982), which prompted the DARPA Strategic Computing Programme in the US and the EEC ESPRIT Programme in Europe (both launched in 1983), followed by similar EEC projects in telecommunications (RACE) and manufacturing (BRITE). Then it was the US Strategic Defense Initiative (SDI) of 1984 that led the European governments to start the Eureka Programme in 1985 and the Japanese government to announce the Human Frontier Science Programme. With their large budgets, high technology programmes are a major force shaping the development and applications of new technologies in the areas of microelectronics, computers, telecommunications, materials, biotechnology. By concentrating large R&D resources and developing major innovations they have a unique influence on economic systems and on the industrial changes of advanced countries.

The most interesting characteristic of high technology programmes is that they combine the forces of industry, scientists and government in a strategy for technological competition between the US, Japan and the EEC. They appear to be the most visible and well-defined attempt to move the research frontier in a particular direction by efforts to improve the technological position, industrial competitiveness and international power of the three most advanced areas. Their nature as tools for competition at the technological frontier is confirmed by the very forms of their development. After one country starts a major high technology programme in a particular direction, the others quickly follow with initiatives of their own, either replicating research in the same area, as in the case of advanced computers, or aiming their technological efforts in a somewhat different direction.

In Europe, the EEC technology programmes, from ESPRIT to BRITE, aim at strengthening its competitive position vis-à-vis the US and Japan, but they have also contributed, by their very design, to greater cooperation among EEC countries. The pooling of resources among firms from different EEC countries, the involvement of research institutes and the matching of the funds by the European Commission have stimulated Europe-wide exchanges and links in innovative activities. On the one hand, these programmes have tried to influence the direction and priorities of the international technology agreements among firms; on the other, they have further supported the growth of cooperative projects among different agents, including firms, universities and government agencies.

Greater opportunities for international cooperation for other agents, including smaller firms and research institutions, and the extension of the EEC programmes also to other fields may further improve the prospects for European cooperation in technology. In fact, the vast majority of research resources in Europe are still mobilized by individual countries, and the funds for EEC R&D programmes amount only to about 3% of the total R&D expenditure of the member countries.

The International Collaboration of Scientists

The pattern of growing international cooperation is found not only in production and technology-oriented activities, but also in science. The nature of scientific research and the *modus operandi* of the scientific community follow their own patterns, which cannot be reduced to economic motivations. In general, international collaboration and exchange among scientific communities has always been more open than in technology or industry, where proprietary knowledge and industrial secrecy restrict the dissemination of information. However, in the recent development of scientific activity a parallel growth of international collaboration has been found. In particular, several studies looking at co-authorship of scientific papers by researchers from different countries have found that international collaboration on common research projects has substantially increased.

Comparing 1973 and 1984 data, a study by Frame and Narin (1988), using the same database on scientific publications as is used in this report and described in Chapter 3, found a doubling of international co-authorship for all countries, with the fastest growth shown by France and Germany. However, countries differ strongly in the level of international collaboration. Only 9.3% of all US papers in 1984 have a foreign co-author, the corresponding figure for British papers being 16.1%, for German papers 18.5%, and for French papers 19.2%. Japan is a remarkable exception, with only 6.8% of its papers involving international collaboration (id. p. 208).[11]

Using the same database for the 1981–86 period, another study (Luukkonen et al. (1992)) has shown that the number of internationally co-authored papers is inversely related to the size of the scientific community of each country. As one might expect, countries with a small scientific community are more likely to find partners in other countries. Looking at the regional distribution of scientific collaboration, the existence of bilateral preferences in the selection of partners has been documented. Geographical distance, language, and political links are all factors which play an important role in defining the pattern of internationally co-authored papers. The study finds that EEC countries seem to be divided into three main clusters. The first represents the German-speaking area (Germany, Austria, Switzerland and the Netherlands). The second includes the Latin countries (France, Italy, Spain and Belgium). In the third, the UK is part of the large English-speaking cluster extending beyond Europe in the United States, Canada and Australia. This evidence may suggest a lack of explicit identity of the European scientific communities. This problem has partially been tackled by specific EEC policies to encourage intra-European cooperation (for their assessment, see Narin and Whitlow (1990)). Other regional clusters of intense scientific collaboration can be found in the European Nordic countries and among Eastern European countries. Outside Europe, the United States concentrates its scientific collaborations largely within the large English-speaking cluster, while Japan has an intermediate position, cooperating both with the US and with other Pacific countries, such as New Zealand, Australia and China.

Patterns of international collaboration among scientific communities are the result of several factors. The high international mobility of researchers working abroad as part of their training and activity is clearly playing a major role. Institutional aspects are also important. First, the procedures of some forms of research funding favour international exchange and cooperation, a variety of EEC programmes having again been developed for this purpose. Second, the need to share the cost of large scientific infrastructures among various countries is an additional factor favouring greater cooperation. However, scientists may also be subject to pressure for greater cooperation similar to that previously described in the case of firms. As research fields become more specialized, international collaboration becomes more important for carrying out world class projects.

2.5. Overview

This Chapter has outlined the main developments in science and technology taking place in advanced countries in the 1980s. S&T activities are becoming more international in scope. For different purposes and through different channels, firms, governments and the academic community are increasingly acting on a global scale, and it is not difficult to predict that this trend will continue in the next decade. This result is not surprising: international integration is growing not only in S&T, but in economic, social, and cultural life as well. Still, in S&T the process of internationalization now appears as a crucial driving force.

The results of the changing innovation systems, and the strategies pursued by governments and firms can be summarized, at the end of this Chapter, in terms of three main processes shaping the current science and technology landscape:

i) Knowledge is becoming more complex and differentiated. Individual firms, research institutions and even countries find it increasingly difficult to cover the entire spectrum of know-how relevant for their activity. As the growing internationalization of science and technology changes the scale of activity, all the agents are under pressure to identify the fields where they should concentrate their efforts to excel at the international level. This increases the incentive to spread research results and exploit innovations in global markets, as well as the need to acquire knowledge developed elsewhere; all this further strengthens the process of internationalization.

ii) Firms cooperate more, sharing their technical knowledge with one another and with other agents at the national and international level. Although greater cooperation has resulted, in some cases, in mergers and take-overs, thus increasing the degree of concentration of some industrial sectors, it is generally taking place in a highly competitive context, favoured by the very process of internationalization.

iii) Nations maintain the distinct nature of their innovation system, providing firms with selected comparative advantages. In the context of greater internationalization and competition, government science and technology policies aim at improving the relative position of the country, its economic performance and international power.

The processes described above offer an overview of the current changes of S&T systems and of the international strategies of firms and governments in these fields. A number of key issues have been identified and a few questions raised. In the following Chapters, detailed evidence of these patterns will be provided by empirical analysis.

Notes

1. See Nelson (1981), Baumol et al. (1989), for productivity; Freeman and Soete (1987), for employment; Soete (1981), Hughes (1986), Fagerberg (1983), for international competitiveness; Denison (1967), Fagerberg (1987), for growth rates.
2. This approach to the analysis of technology has been developed, among others, by Rosenberg (1976, 1982), Nelson and Winter (1982), Freeman (1982), Pavitt (1984), Dosi et al. (1988).
3. A study of the world's 800 largest firms, which account for 90% of world trade, estimated that 34% of all trade is intra-firm. This share increases to 43% for the firms with a higher research intensity, and falls to 13% for the low-technology industries (see *The Economist*, 1st March 1986, p. 61). The case of the US exemplifies these processes. Foreign affiliates of US multinational corporations are now responsible for about 20% of all US imports: two thirds of the US trade deficit would disappear if US firms produced at home what they import from their foreign subsidiaries (Faux, (1988)).
4. Studies of the accelerating pace of technology transfer within firms include Mansfield et al. (1982), and Vernon (1981).
5. More recent studies of the international integration of large firms' innovative activities can be found in Cantwell (1989) and Cantwell and Hodson (1990).
6. A study by the US Office of Technology Assessment based on OECD data found that the rate of import penetration in R&D intensive industries in the US doubled between 1975 and 1980, going from 7.7 to 14.3 per cent. In Germany it increased from 28.4 to 42.5 per cent; in France from 23.3 to 28.8 per cent; in the UK from 31.2 to 44.2 per cent; in Italy from 25.3 to 33.4 per cent. Only in Japan was the increase limited: from 6.4 to 7.9 per cent (Office of Technology Assessment 1988, p. 322).
7. A major international comparative study on national innovation systems has been carried out by Nelson and associates (Nelson, forthcoming).
8. See, among others, Baumol et al. (1989). The process of convergence does not, however, include many countries besides the US, Japan, and the most advanced European countries. Greater international technology flows are a necessary but not a sufficient condition for allowing less advanced countries to reach the technology frontier. The conditions which have made it possibile for Europe and Japan to "catch up" are not easily found elsewhere, with the possible exception of a few newly industrial countries of the Pacific. On the contrary, there is some evidence suggesting an increasing technology "gap" between the more advanced and the developing countries.
9. An example is provided by the new emerging strengths of Japan also in basic research. A 1986 report of the US National Security Council found a Japanese superiority in R&D for semiconductors, industrial automation, computer architecture and telecommunications components. The report concluded that "the conventional model of US technological leadership in basic research followed by more successful Japanese commercial exploitation is no longer accurate in many of the critical technologies targeted by the Japanese" (quoted in Reich 1987, p. 65). The effect of R&D on US-Japanese trade is explored in Audretsch and Yamawaki, 1988.
10. A 1985 OECD report stated that firms "have not previously cooperated so directly, on such a scale, in planning, financing and carrying out joint R&D" (OECD, 1985, p. 64). Several studies have examined the new pattern of international cooperation among firms in particular fields; see Chesnais (1988), Van Tulder and Junne (1988), Finan et al. (1986), OECD (1985b), Hagedoorn and Schakenraad (1992b).
11. An additional study of co-authorship patterns can be found in Miquel et al. (1989).

CHAPTER 3

An Overview of Scientific and Technological Activities in the Advanced Countries

3.1. The Indicators of S&T Activities

The previous Chapters have outlined the processes of growing international competition and cooperation which are transforming the structure, performance and hierarchy of S&T activities in the advanced countries. The position of individual countries in these processes varies widely, both the strategies they pursue and the results they achieve being strongly influenced by the quantity, nature, and intensity of the technological knowledge and skills accumulated in the past. The differences across countries require a quantitative investigation of the characteristics shown by their innovative activities. This is the basis for comparison of national performances and the technology policies developed as a response to the competitive challenges in international markets.

Comparative analyses of national technological capabilities have been carried out over the post-war period by considering data on productivity, growth rates, structure of international trade, and industrial organization (see, among a large literature, Denison (1967), Maddison (1987), Baumol et al. (1989)). Common to these works is the crucial role attributed to technology as a determinant of economic performance. In these studies the progressive narrowing of the "technology gap" between the United States and the other industrialized countries is analyzed in detail. In these studies, however, technological change is viewed essentially as embodied in such economic variables as trade or production. This book employs a different methodology by considering S&T capabilities in their own right. The indicators used focus on technological change as disembodied from economic variables (former works in this field include Patel and Pavitt (1987a), and Soete (1987)), which allows us to examine the specific dynamics of technology across countries, sectors, and over time.

The following Chapters of this report examine national patterns of innovation, with particular emphasis on sectoral specialization. Since technological change is a highly diversified process across sectors, only a disaggregated analysis can provide an adequate picture of the developments in each country. A preliminary analysis of the aggregate level of S&T capabilities in the advanced countries is however required. The inventive and innovative activities of each country in any specific field are related to the magnitude of national efforts in science and technology. Nations with limited resources devoted to S&T might be forced to concentrate on certain areas rather than on others. This Chapter investigates the amount and distribution of the resources devoted to S&T by country. In more specific terms, the following points are considered:

– the distribution of resources devoted to S&T across advanced countries;
– the S&T intensity of their economies. More particularly, the process of convergence among advanced countries will be examined;
– the distribution of S&T within each country between the public and the business sectors, and among "scientific" and "technological" activities;
– the main trends over time.

This aggregate analysis will set the context for the sectoral results discussed in subsequent Chapters. It will also provide some fresh evidence on the dynamics of S&T and on the changes in the hierarchy of countries. Two main elements emerge from the evidence available, namely the progressive narrowing of the technology gap between the US and the other industrial countries, and the dramatic rise of Japan as an economic and technological power (see, among a large literature, Nelson (1989), Pianta (1988b)).

Diverging signals may be observed in the performance of European countries. Recent patterns have shown that the fears voiced in 1960s of an increasing technology gap between Europe and the US (Servan-Schreiber, 1968) were unfounded. On the contrary, optimistic views about European performance have emerged (Thurow, 1985), although a consensus on the performance and prospects of European technology has not been reached. It is not surprising that the assessment of European technological capabilities has often lead to diverging opinions. Europe is the sum of various countries, each with its own particular history, capacity and performance, which are not easily combined into a single unit. Moreover, firms, national governments and European institutions have developed a variety of policies, sometimes with diverging objectives. An adequate analysis of European technological perspectives must therefore start from detailed documentation at the level of country and sector.

The S&T indicators relevant for the analysis carried out in this book – including resources devoted to R&D, patenting, and bibliometric indicators – will be used at the aggregate level in this Chapter, and at the sectoral level in Chapters 4 to 7. None of these indicators, and not even all the S&T indicators taken together, are able to account fully for the richness and variety of inventive and innovative phenomena. They provide, nevertheless, a better understanding of reality than anecdotal and fragmentary information. Table 3.1 lists the main advantages and disadvantages of the indicators used in this report.

An indicator is useful insofar as it is a fair approximation of time-series and/or cross-sectional variations in the phenomena considered. Though an indicator may overestimate or underestimate a phenomenon, if this over (or under) estimation is systematic and fairly stable across sectors and over time, it is still useful for understanding differences across areas and periods. Indicators are a sort of "tip of the iceberg" of innovative activities; as long as there is a stable relationship

Table 3.1
PROXY MEASURES OF SCIENTIFIC AND TECHNOLOGICAL ACTIVITY

Measure	Strengths	Weaknesses
Research and Development	* Regular data collection * Sectoral uniformity across industries * Internationally comparable	* Monetary adjustments required for international comparability * Underestimates innovation in small firms * Excludes design, software and production engineering
Patenting	* Regular data collection * Detailed break-down for technological fields * Internationally comparable * Direct measure of technological output	* National systems are biased by domestic inventors * Not all inventions are patented * Not all inventions are patentable * Differences in the propensity to patent across sectors
Patent citations	* Measure of impact of patented inventions	* Differences in the number of citations across time and sectors
Scientific papers	* Detailed break-down for scientific disciplines * Internationally comparable * Direct measure of scientific output	* Databases include a sub-set of world publications * Differences in the propensity to publish across disciplines * Language barriers
Scientific citations	* Measure of impact of scientific literature	* Differences in the number of citations across time and sectors

between the visible and the hidden part of the "iceberg", they make it possible to measure the relevant aspects of technological change.

3.2. The Resources Devoted to Research and Development

The resources devoted to formal R&D activities have been, for more than a quarter of a century, the most widely used indicator of scientific and technological activities. In recent years, however, increased criticism of their value as a technological indicator has emerged. It is argued that many innovations are not produced within the R&D laboratories but on an informal basis by entrepreneurs, technicians, engineers, and workers (see, among others, Archibugi et al. (1987), OECD (1990)). We do not deny the importance of technology other than formalized R&D, but still maintain that the latter represents one of the most relevant inputs for invention and innovation. The resources devoted to R&D remain one of the most informative indicators of the comparative effort in S&T at the country level.

Table 3.2 shows the percentage distribution of R&D expenditure in the OECD area broken down by country for the years 1975, 1981 and 1987.[1] The share of the largest R&D spender of the OECD area, the US, is greater than that of the EEC and Japan together, in spite of having decreased slightly over the period considered. The EEC reduced its share from 31.4% to 27.9% of total OECD R&D, while Japan[2] increased its share from 12.2% to 16.2%. Within the EEC, the three largest R&D spenders, West Germany[3], France and the United Kingdom, decreased their share, and a similar drop is found for the Netherlands and Belgium. Italy and Spain, on the contrary, increased theirs, followed also by Portugal and Greece. Outside the EEC, Sweden, Norway, Finland and Australia show an increasing share of OECD R&D expenditure, while Switzerland is losing ground.

The ratio between R&D expenditure and Gross Domestic Product, shown in Table 3.3 for selected years, is a widely used indicator of the S&T intensity of each country. A small number of OECD countries, including the three largest R&D spenders – the United States, Japan, and West Germany – are now above 2.5% of GDP. The EEC as a whole has an R&D intensity substantially lower than the other two major industrial areas, equal to 1.9% of GDP. Within the European Community, a wide range of R&D intensities can be found. Only West Germany has an R&D intensity comparable to the US and slightly above Japan. The two other largest R&D spenders, France and the United Kingdom, devote about 2.3% of GDP to R&D, as do the Netherlands. Belgium, Denmark, and Italy spend between 1 and 2% of their GDP on R&D. Ireland, Spain, Portugal and Greece devote less than 1% of their GDP to R&D. Two other small sized European countries, Sweden and Switzerland, emerge among the most R&D-intensive economies of the OECD area.

An important qualification of R&D data is provided by the ratio of civilian R&D to GDP in 1987, shown in the last column of Table 3.3. As several studies have shown, military R&D is less likely to produce benefits in terms of national competitiveness and economic performance[4]. The R&D intensity of the US is considerably lower if military R&D is excluded, and the same holds, to a lesser degree, also for the United Kingdom and France. The highest civilian R&D intensity

TABLE 3.2

GROSS DOMESTIC EXPENDITURE ON RESEARCH AND DEVELOPMENT IN OECD COUNTRIES

Percentage distribution within the OECD in 1975, 1981, 1987
and R&D expenditure in 1988 in million US dollars at 1988 prices and purchasing power parities exchange rates

Countries	Percentage distribution			Million current ppp $
	1975	1981	1987	1988
United States	49.63%	48.27%	49.02%	137,816
Japan (1)	12.18%	14.62%	16.22%	47,649
EEC 12	31.36%	30.22%	27.91%	78,950
W. Germany	9.64%	9.46%	8.75%	24,626
France	6.68%	6.70%	6.16%	17,528
United Kingdom	8.17%	7.54%	6.10%	16,184 (2)
Italy	2.80%	2.85%	3.14%	10,004
Netherlands	2.02%	1.63%	1.57%	4,365
Belgium	0.85%	0.85%	0.74%	2,056
Denmark	0.36%	0.34%	0.37%	973 (2)
Spain	0.65%	0.61%	0.79%	2,437
Ireland	0.10%	0.09%	0.10%	270
Portugal	0.08%	0.10%	0.10%	255 (3)
Greece	0.02%	0.07%	0.09%	252
Switzerland	1.41%	1.14%	1.02%	2,866 (3)
Sweden	1.18%	1.25%	1.34%	3,533 (2)
Austria	0.43%	0.50%	0.44%	1,258
Canada	2.23%	2.37%	2.26%	6,322
Australia	0.92%	0.94%	0.98%	2,625 (2)
Finland	0.28%	0.33%	0.41%	1,204
Norway	0.37%	0.36%	0.40%	1,194 (2)
TOTAL OECD	100.00%	100.00%	100.00%	283,417

Source: CNR-ISRDS elaboration on OECD data, MSTI, April 1990

Missing values have been replaced by the estimated values of the regression line in the 1975-89 period.
(1) Data for Japan are the OECD adjusted values
(2): 1987
(3): 1986

is found in West Germany and Japan among the largest spenders, as well as in Sweden and the Netherlands.

Table 3.4 reports the average annual growth rates of R&D expenditure in real terms. From 1975 to 1989, a remarkable increase occurred in all countries. The

TABLE 3.3

EXPENDITURE ON RESEARCH AND DEVELOPMENT AS A PERCENTAGE OF GDP

Gross domestic R&D expenditure and estimates of civilian R&D expenditure as a percentage of GDP

Countries	Total R&D expenditure, % of GDP				Civil R&D, % of GDP
	1975	1981	1987	1989	1987
United States	2.32	2.45	2.90	2.80	2.1
Japan (1)	1.81	2.14	2.67	n.a.	2.6
EEC 12	1.59	1.71	1.95	1.95	1.7 (9)
W. Germany	2.24	2.42	2.83	2.85	2.7 (8)
France	1.79	1.97	2.29	2.33	1.9
Un. Kingdom	2.18	2.42	2.27	n.a.	1.8 (7)
Italy	0.84	0.87	1.19	1.25	1.1 (8)
Netherlands	2.02	1.88	2.33	2.30 (6)	2.3
Belgium	1.30	1.31 (3)	1.65	1.61 (6)	1.6
Denmark	1.01	1.10	1.43	n.a.	1.4
Spain	0.35	0.40	0.62	0.72	0.6
Ireland	0.85	0.73	0.94	0.94 (6)	0.9
Portugal	0.27 (2)	0.35 (4)	0.45 (5)	n.a.	n.a.
Greece	n.a.	0.21	0.33 (5)	0.37 (6)	0.3 (8)
Switzerland	2.40	2.29	2.88 (5)	n.a.	n.a.
Sweden	1.80	2.30	2.99	n.a.	2.8
Austria	0.92	1.17	1.32	1.32	1.3
Canada	1.10	1.21	1.36	1.29	1.3
Australia	1.00 (2)	1.00	1.19	n.a.	1.1 (5)
Finland	0.91	1.19	1.73	1.80	1.7
Norway	1.34	1.29	1.81	1.87	1.7

Source: CNR-ISRDS elaboration on OECD data, MSTI, April 1990

(1): Data for Japan are the OECD adjusted values
(2): 1976 (5): 1986
(3): 1979 (6): 1988
(4): 1982 (7): 1985
(8): Value estimated as the share in GDP of GERD less government appropriations for defence R&D
(9): Estimate
n.a.: not available

growth rate achieved by Japan (7.4%) is substantially higher than that of the US (4.6%) and of the EEC (4.1%). The slowest growth occurred in European countries such as the UK, the Netherlands and Switzerland. Conversely, the fastest growth rates were achieved, with the notable exception of Japan, by economies with low

TABLE 3.4

RATES OF GROWTH OF R&D EXPENDITURE IN OECD COUNTRIES

Average annual percent growth in real terms
(using OECD implicit GDP price indexes)

Countries	1975-78	1979-82	1983-85	1986-89	1975-89
United States	3.71	4.46	9.55	2.34	4.58
Japan	4.62	9.28	9.81	7.41 (11)	7.42 (15)
EEC 12 (1)	2.89	4.38	5.86	4.23	4.05
W. Germany	2.06 (2)	2.07 (4)	6.20	4.38 (14)	4.25
France	2.48	6.92	4.92	4.31	4.43
United Kingdom	3.14	3.04 (5)	3.59	1.23 (12)	2.61 (16)
Italy	-0.06	8.82	11.74	6.60	6.10
Netherlands	0.63	1.28	4.66	3.70 (11)	2.87 (15)
Belgium	4.24 (2)	5.99 (6)	3.49	2.20 (11)	3.72 (15)
Denmark	1.91	6.53	6.97	7.37 (12)	5.39 (16)
Spain	2.38	8.19	10.46	12.05	7.76
Ireland	0.50 (2)	4.51	12.41	6.48 (11)	4.34 (15)
Portugal	12.81 (3)	4.98 (7)	5.54 (10)	n.a.	7.27 (17)
Greece	n.a.	5.79	17.05	7.30 (11)	n.a.
Switzerland	0.75	0.46 (4)	11.15 (8)	n.a.	3.35 (17)
Sweden	2.22 (2)	12.32 (4)	9.39	4.48 (13)	6.20 (16)
Austria	n.a.	n.a.	3.52	3.42	5.21
Canada	3.93	8.66	8.88	0.23	4.66
Australia	1.88 (3)	3.32 (5)	8.61 (9)	0.74 (12)	4.79 (18)
Finland	4.77 (2)	10.09 (4)	11.34	7.19	8.32
Norway	6.43	4.02	13.00	3.31 (14)	5.41

Source: CNR-ISRDS elaboration on OECD data, MSTI, April 1990

(1): Missing values have been replaced by the estimated values of the regression line in the 1975-89 period

(2) 1975-77	(7) 1978-83	(12) 1986-87	(17) 1975-86
(3) 1976-77	(8) 1983-86	(13) 1985-87	(18) 1976-87
(4) 1979-81	(9) 1984-86	(14) 1985-89	
(5) 1978-81	(10) 1981-84	(15) 1975-88	
(6) 1979-83	(11) 1986-88	(16) 1975-87	

n.a.: not available

R&D intensity, such as Spain, Portugal and Italy. Since countries with a low R&D intensity have generally grown faster than the average, the variability of R&D intensity across OECD economies was somewhat reduced.

The same Table shows the growth rates of four subperiods. The highest average annual rates of increase in R&D spending were achieved in the 1983–85 period, while in more recent years a slowing down of growth rates is evident; in some cases resources devoted to R&D increased at a slower rate than GDP. A period of slower growth in R&D spending seems to be opening, in particular in the US, the UK, Canada and Australia (the same concern was recently expressed by the National Science Board, 1989). If confirmed, this trend would have major implications for S&T policy; the search for innovation might shift from budgeting for ever increasing R&D resources to more selective efforts to improve the national competitive position. Several aspects may contribute to such a slowdown. The returns from R&D spending may be decreasing above a certain threshold, especially in the business sector. Also, a sort of "R&D saturation effect" may operate in the most advanced countries, as experience shows that no country has yet devoted more than 3% of its GDP to R&D. The R&D slowdown will, however, need to be investigated in greater detail in both theoretical and empirical terms.

As the quantitative growth of resources devoted to R&D slows down, it becomes crucial to identify the fields where competition for innovation is taking place, and a few explanatory hypotheses might be attempted:

a) innovative activities may be located not only in R&D laboratories but also in the wider economic and social systems involving education, managerial skills, and diffusion of knowledge. In other words, innovative activities *outside* the R&D laboratories may be growing at the expenses of those carried out *inside* them;

b) international agreements among firms and public research institutions and takeovers of foreign companies may act as a substitute for greater domestic R&D expenditure;

c) consequently, each country is likely to try to appropriate the results of its innovations in external markets, or to move from "more R&D" to "R&D more exploitable at the global scale".

The last two hypotheses would suggest that countries are concentrating their efforts in the sectors of already existing strengths, while relying on international trade and partemship in their sectors of weaknesses. These questions will be examined in the following Chapters.

One of the main factors qualifying the structure of national innovative systems is the financing and performance of R&D in the public and business sectors. Table 3.5 provides a breakdown by source of funds and sector of performance for OECD countries in 1987. The different contributions of the public and business sectors influence the outcome of R&D activities. Countries where companies fund a large part of R&D are expected to produce research results which are more directly related to the companies' activity and are more likely to be proprietary in nature. The R&D funded by the public sector, on the other hand, is generally aimed at improving national economic performance and at furthering the advancement of knowledge. Countries with a large share of R&D funded by and performed in public institutions are more likely to produce a lower amount of innovations directly linked to market activities.

In countries such as Switzerland, Japan and West Germany, the business sector contributes more than half of the national R&D resources, but in the majority of countries its contribution is lower. In the US, the business sector funds half the total R&D, but performs more than 70%, the difference being due to government funded military and space programmes. A similar pattern is followed in Europe by France. The share of foreign funded R&D is particularly high in the UK and Canada, where it is close to 10%, reflecting the importance of research laboratories established by foreign firms. The following sections will explore how these differences in the sources of finance and of performance are reflected in the indicators of S&T output.

TABLE 3.5

EXPENDITURE ON RESEARCH AND DEVELOPMENT BY SECTOR OF FINANCING AND SECTOR OF EXECUTION - 1987

	% by sector of performance					% by source of funds			
Countries	Business enterprise	Government	Higher Education	Private non-profit	Total	Business enterprise	Government	Other national	From abroad
United States	72.6	10.6	14.1	2.7	100	49.5	48.7	1.8	0.0
Japan (1)	70.9	10.3	14.0	4.8	100	73.5	19.6	6.8	0.1
W. Germany	72.2	12.7	14.6	0.5	100	63.6	34.7	0.4	1.3
France	58.9	25.2	15.0	0.9	100	41.8	51.7	0.6	5.9
United Kingdom	67.0	15.1	14.2	3.7	100	49.7	38.7	2.5	9.2
Italy	57.2	22.6	20.2	n.a.	100	41.7	54.0	n.a.	4.3
Netherlands	59.2	17.3	21.4	2.1	100	51.8	44.3	2.0	2.0
Belgium	72.9	4.4	18.9	3.8	100	70.7	27.6	0.7	1.1
Denmark	55.6	19.4	23.9	1.1	100	48.7	45.9	2.8	2.7
Spain	57.3	26.3	15.5	0.9	100	48.8	48.5	1.1	1.5
Ireland (2)	53.6	24.3	20.8	1.3	100	48.0	43.8	1.5	6.6
Portugal	26.3	36.0	30.1	7.6	100	26.8	63.5	6.7	2.9
Greece (2)	28.6	49.8	21.6	n.a.	100	23.2	74.4	n.a.	2.4
Switzerland	77.7	6.3	12.8	3.2	100	78.9	21.1	n.a.	n.a.
Sweden	66.8	4.2	28.9	0.1	100	60.0	36.9	1.5	1.6
Austria	n.a.	n.a.	n.a.	n.a.	100	48.8	48.5	0.3	2.4
Canada	55.0	20.4	23.3	1.3	100	41.7	41.7	3.9	9.2
Australia(2)	36.5	34.9	27.1	1.5	100	35.9	60.7	2.5	0.9
Finland	58.9	20.1	20.6	0.3	100	58.8	38.9	1.2	1.0
Norway	62.0	15.8	21.2	1.0	100	50.3	46.8	1.1	1.7

Source: CNR-ISRDS elaboration on OECD data, MSTI, April 1990

(1): Data for Japan are the OECD adjusted values
(2): 1986
n.a.: not available

3.3. Patents

Patenting will be extensively used in this report as an indicator of technology. Patents are the outcome of S&T activities of a proprietary nature and are likely to generate business applications. In other words, patents are more likely to reflect technological than scientific activities. Like any other indicator, patenting has its own strengths and limitations. Suffice it to highlight what patents can and cannot measure[5].

TABLE 3.6

PATENTS REGISTERED BY ADVANCED COUNTRIES IN THE MAIN PATENTING INSTITUTIONS

Percentage distribution of patents registered at the European Patent Office, Japan, W. Germany, France and in the US by country of origin

Countries	EPO 1980-89 applic.	Japan 1979-87 applic.	Japan 1979-87 granted	W. Germany 1982-87 applic.	W. Germany 1982-87 granted	France 1981-87 applic.	France 1981-87 granted	United States 1979-88 granted	United States 1979-88 citations
United States	25.82%	4.47%	7.86%	6.48%	13.71%	10.79%	15.54%	56.31%	59.86%
Japan	15.57%	89.42%	83.67%	9.48%	13.85%	7.36%	9.40%	16.26%	17.46%
EEC 12	47.64%	4.50%	6.09%	75.75%	63.03%	74.24%	62.50%	19.47%	16.72%
W. Germany	23.26%	2.28%	2.98%	71.00%	52.74%	10.58%	15.54%	9.41%	7.92%
France	8.98%	0.71%	0.99%	1.30%	3.61%	55.88%	36.03%	3.36%	2.81%
United Kingdom	7.26%	0.68%	0.90%	1.14%	2.16%	1.91%	3.49%	3.55%	3.41%
Italy	3.27%	0.27%	0.30%	1.25%	1.33%	3.10%	3.31%	1.31%	0.96%
Netherlands	3.22%	0.40%	0.70%	0.59%	1.98%	1.01%	2.58%	1.07%	1.01%
Belgium	0.90%	0.06%	0.10%	0.15%	0.29%	0.59%	0.66%	0.37%	0.34%
Denmark	0.33%	0.05%	0.09%	0.19%	0.60%	0.36%	0.31%	0.23%	0.17%
Spain	0.27%	0.03%	0.03%	0.11%	0.28%	0.76%	0.55%	0.12%	0.06%
Ireland	0.09%	(1)	(1)	0.01%	0.01%	0.03%	0.02%	0.04%	0.02%
Portugal	0.01%	(1)	(1)	(1)	(1)	0.01%	0.01%	0.01%	0.01%
Greece	0.03%	(1)	(1)	0.01%	0.03%	0.01%	0.01%	0.01%	0.01%
Switzerland	4.46%	0.52%	0.82%	2.19%	3.61%	2.05%	3.32%	1.81%	1.50%
Sweden	1.63%	0.23%	0.32%	0.67%	1.28%	0.76%	1.58%	1.16%	0.92%
Austria	1.07%	0.08%	0.12%	0.57%	1.83%	0.60%	0.49%	0.44%	n.a.
Canada	0.91%	0.11%	0.17%	0.15%	0.33%	0.29%	0.42%	1.83%	1.50%
Australia	0.57%	0.10%	0.08%	0.07%	0.34%	0.25%	0.23%	0.47%	n.a.
EPO countries	54.66%	-	-	-	-	-	-	-	-
Others	2.35%	0.59%	0.87%	5.06%	2.21%	3.66%	6.51%	2.24%	2.04%
WORLD	100%	100%	100%	100%	100%	100%	100%	100%	100%

Sources: CNR-ISRDS elaboration on WIPO and EPO data
(1): Less than 0.00%.
n.a.: not available

a) Patents represent the outcome of the inventive process, and more specifically of those inventions which are expected to have a business impact.

b) Patents are registered on a regular basis by specific offices in the majority of countries. Patent statistics started to be collected a couple of centuries ago for legal purposes and are today the technological indicator with the longest time series, providing unique historical information about innovation.

c) Obtaining patent protection is time consuming and costly. It is likely that applications are presented for those innovations which, on average, are expected to provide such benefits as to compensate the costs involved in obtaining patent protection.

d) Patent applications are accurately broken down by technological classes by independent external examiners. This provides internationally comparable information on sectoral patenting up to the five digit level of disaggregation. Since the following Chapters are devoted to a sectoral analysis of technological change, the detailed classifications provided by patent statistics have a strong advantage over other indicators.

However, we are not unaware of the limitations of patenting as a technology indicator.

a) Not all inventions are patented. Firms often protect their innovations with other methods[6], and most notably by recourse to industrial secrecy or to other methods of appropiation.

b) Not all inventions are technically patentable. This is the case of software, which is coming to play an increasingly important role in current technological advance.

c) The propensity to patent (i.e. the number of patents registered for each unit of inventive and innovative activity, see Scherer, 1983) varies greatly across technological areas. While in pharmaceuticals, for example, a large part of the outcome of inventions is codified in patent applications, in other equally S&T intensive fields, such as nuclear technology, only a minority of inventions are actually patented.

d) Firms have a different propensity to patent in each national market, according to their business expectations on the possibility of exploiting their inventions. The size of national markets and the level of integration in international trade affects the number of foreign patent applications received by each country. Moreover, each patent office receives an over-represented number of applications by domestic inventors and firms.

e) In spite of the international patent agreements among the majority of industrial countries, each national patent office has its own institutional characteristics; the attractiveness for applicants of any patent institution depends on the nature, costs, length and effectiveness of the protection accorded.

Table 3.6 reports the share of patents registered during the 1980s in five patent offices: the US, Japan, France, West Germany, and the European Patent Office, according to the country of origin of the applicant. Patent data are aggregated in periods spanning several years in order to avoid random fluctuations.

Each patent office receives applications from both national and foreign inventors, although the proportion varies significantly from country to country. Countries with low S&T budgets generally have a restricted internal market and are not particularly attractive for foreign inventors. In large countries there are a great number of foreign applications. The institutions which receive a large number of foreign patents are also those which are better placed to provide international comparisons on the technological activities of the different countries.

The technical importance of patents and their economic impact varies widely and the absolute numbers of patents registered do not provide entirely reliable information on the quality and impact of patented innovations. It is well known that the quality and impact of patents are highly skewed; while few patents have an important economic impact, many of them are never used. Several attempts have been made to obtain information on the quality and impact of individual patents. For example, Schankerman and Pakes (1986) take into account the money spent on renewal fees in order to assess the value of each patent. Another method of obtaining information on the impact of a patent in technological evolution is to consider the number of times a patent is cited in subsequent patents. An often cited patented invention is likely to be associated with important innovations. Data on patent citations are available for the US and refer to the list of citations in the front page of the patent prepared by the US Patent Office examiner. This external assessment of the links between a new patent and previous ones ensures a fairly standard approach to the use of citations and is much more reliable than the use of the citations listed by each inventor, which may both overstate the importance of previous patents registered by the same inventor or firm, and ignore other inventors' patents.[7]

The older the year of registration of a patent, the greater the probability of its being cited in subsequent patents. On average, therefore, old patents receive more citations than recent ones. The distribution over time of citation activity may differ across countries and sectors, and some caution is needed when citations referring to patents of different years are grouped together. However, when data for several years are grouped together, the share of citations received by the patents of advanced countries provides a reliable indicator of the overall impact of the inventions of each country. The share of citations reflects the countries' level of accumulated knowledge.

Among the different patent data listed in Table 3.6, the first and most important database to be considered is the one on patents granted in the US. The US is the largest and technologically more important market of the world, and it is reasonable to expect that the more relevant and valuable inventions would be patented (also) in the US. The Table shows for each country the share of patents granted and the share of patent citations received.[8] More than 45% of patents in the US were granted to foreigners. In absolute numbers, the US patent office has received by far the largest number of foreign applications. Trends over time are also instructive: the absolute number and the share of foreign patents have steadily grown over the last twenty-five years, reflecting the internationalization of markets as well as the narrowing of the technology gap between the US and the other industrialized countries.

In the US, Japan has registered almost as many patents as all the 12 EEC countries together, and in terms of citations its share is higher than the EEC. It should be noted that Japanese patents have received a higher average number of citations than any other country, US included. Within the EEC, the largest share of patents in the US is held by West Germany with 9.4%, although it falls to 7.9% in terms of patent citations. Great Britain and France have respectively 3.6% and 3.4% of all patents, while the share of citations is 3.4% for the former and 2.8% for the latter. Switzerland has a higher number of patents than Italy and a higher impact. Small countries like the Netherlands and Sweden have a share above 1%.

The second patent institution considered in this report is the European Patent

Office, founded in 1978 on the basis of an agreement among 13 European countries: Austria, Belgium, France, West Germany, Greece, Italy, Liechtenstein, Luxemburg, the Netherlands, Spain, Sweden, Switzerland and the United Kingdom. Nine of these countries are also EEC members. This patent office is the only one in the world to be truly international, since a single application can potentially be extended to all the 13 member countries.

Total applications at the EPO have grown from 11,000 in 1979 to nearly 50,000 in 1989. Less than half (47.6%) of the applications come from EEC countries, while 54.7% come from EPO member countries.[9] The US has the largest share of patents, equal to 25.8%. Although the US ranks first, its share of patents in the European market may underestimate the US technological potential as measured by its share of either R&D or other indicators. The share of West Germany at the EPO (equal to 23.3%) is close to that of the US. Japan ranks third, with a share equal to 15.6%, much higher than France (9.0%) and the UK (7.3%).

The share of patents held by European countries at the EPO is obviously larger than their share in the US. However, the ranking in both institutions is very similar, as it depends on the overall size of a country's inventive activities. The most notable exceptions are the UK, whose share of patents is greater than France in the US, but smaller at the EPO, and Sweden, which ranks 6th among European countries in the US, but is overtaken by the Netherlands at the EPO. Both the UK and Sweden are countries with intense trade and cultural links with the US.

The third database considered here is the Japanese Patent Office. The institutional nature of this patent office is substantially different from those of the other advanced countries, since it asks inventors to split in several applications the different technical aspects related to the same invention. It also differs in the procedures adopted to evaluate applications, as indicated by the fact that only one third of the applications are granted a patent.[10] Applications to the Japanese patent system come mainly from national inventors and firms, which account for nearly 90% of the applications and for over 83% of the patents granted. This figure shows the extent to which the Japanese patent system is oriented towards domestic inventions, a fact which has led to its exclusion from the sectoral analysis carried out in Chapter 4.

The next Chapter considers patent applications and patents granted in two other national patent officies, i.e. France and West Germany. In these countries, the average number of patent applications, especially the foreign ones, filed directly at the national offices has been steadily eroded by the growth of the applications to the EPO. However, the EPO does not erode the number of domestic applications, which makes it possible to consider the differences between domestic and foreign patenting for France and West Germany. Table 3.6 shows that domestic inventors account for 70.6% of the applications in West Germany and for 55.9% in France. In both cases the share of patents granted to domestic applicants is considerably lower, being equal to 52.5% in West Germany and 36.0% in France. In France, the US share is higher than the Japanese, while in Germany the share of Japanese patents is higher than that of the US, a result which may stimulate further investigation into the patenting strategies of Japanese companies.

Table 3.7 provides some basic data on the changing pattern of patenting in advanced countries. The average growth rates in domestic patents during the 1980s (i.e. patents registered by residents in their own country) shown in column 1 demonstrate that the number of patent applications did not grow significantly. Eight countries display a negative rate of change and another three countries growth rates below 2%. This pattern of stagnation or decline in domestic patenting should be compared to the rates of change of R&D inputs already shown in the previous section. In spite of the slowdown after 1985, R&D expenditure has grown at a substantially faster rate than domestic patents. Some authors (Evenson, 1984) relate the slowdown in domestic patent applications to decreasing productivity in scientific and technological research. However, the decrease in domestic patents is also likely to be related to greater awareness on the part of applicants of the opportunities offered by the patent system; in fact, the sluggish or negative growth rate is generally associated to a decreasing number of applications presented by individual inventors, which are less likley on average to become actual innovations.

The greater the resources devoted to S&T, the more we would expect firms to try to appropriate the benefits of their innovations in several markets, and in fact the total number of foreign patents (i.e. patent applications filed by foreign applicants in a given country) has significantly increased in all OECD countries (column 2 of Table 3.7). Although applications registered by foreigners in each country have risen substantially, considerable differences may be observed across countries. Many small and medium sized European countries – including Greece, Austria, Spain, etc. – show a two digit growth rate. Markets on the fringes of international competition seem to become the targets for companies expanding their activities, and many firms with a largely domestic orientation will face an increasing challenge from foreign competitors. Japan is again a significant exception: the extensive penetration of Japanese firms in the Western technology markets shown above has not been paralleled by an equally widespread presence of Western companies in the Japanese market. The growth rate in foreign patent applications has been relatively moderate, and far below the growth rate of domestic applications.

A specular information on the internationalization of technology markets is provided by the share of external applications (i.e. the applications presented by firms and inventors of each country in other countries), shown in column 3 of Table 3.7.[11] All countries show a rapidly growing effort to patent abroad, with the aim of appropriating the returns from their inventions also in foreign markets. Among the larger countries, Japan ranks first. The divergent trends in the growth rates of domestic and foreign patent applications suggest that the reduction in the volume of domestic applications has principally affected those inventions with less certain business potential. This is consistent with the findings of a study showing that, although the number of patents granted in selected European countries has dropped, the money spent on renewal fees has not.[12]

Columns 4 and 5 of Table 3.7 present the ratio of external to domestic patents for 1979 and 1988, which shows a substantial rise in all countries. Two different factors have contributed to this: a) a rise in the average number of countries in which each patent extended abroad is registered; and b) an increase in the number of patents which are extended abroad. However, our data do not make it possible to distinguish the relative importance of these two factors. While the trend towards

TABLE 3.7

RATES OF CHANGE IN DOMESTIC, FOREIGN AND EXTERNAL PATENTS IN THE OECD COUNTRIES

Countries	Rates of change in Domestic Patents 1979-88 (1)	Foreign Patents 1979-88 (2)	External Patents 1979-88 (3)	Ratio of external to domestic patents 1979 (4)	1988 (5)
United States	2.44%	6.30%	7.50%	1.73	2.67
Japan	8.30%	3.85%	11.53%	0.25	0.33
EEC 12	1.10%	9.38%	7.49%	2.06	3.90
W. Germany	0.54%	4.87%	6.79%	2.28	3.93
France	1.16%	5.95%	7.64%	2.41	4.22
United Kingdom	0.64%	5.49%	8.34%	1.37	2.65
Italy	-11.59%a	10.01%a	8.50%	1.97b	n.a.
Netherlands	2.00%	10.17%	6.91%	5.18	7.90
Belgium	-1.05%	10.15%	6.38%	3.65	7.00
Denmark	3.28%	7.24%	13.47%	2.74	6.38
Spain	-0.31%	11.77%	5.06%	0.92	1.48
Ireland	8.39%	4.26%	14.12%	0.81	1.29
Portugal	-6.19%	5.33%	29.86%	0.10	1.94
Greece	-0.18%	27.88%	8.96%	0.07	0.15
Switzerland	-2.21%	9.87%	3.38%	4.60	7.58
Sweden	-2.39%	10.12%	8.56%	2.51	6.52
Austria	-0.75%	13.51%	7.44%	1.66	3.39
Canada	6.28%	2.88%	8.31%	2.83	3.35
Australia	3.26%	5.31%	17.36%	0.70	2.22

Source: CNR-ISRDS, elaboration on OECD, Main Science and Technology Indicators Data, 1990

(1) Patents by residents of the country
(2) Patents by foreigners in the country
(3) Patents by residents extended in other countries
(4) (5) Average number of patent applications extended abroad for each domestic patent application.

a. 1980-88
b. 1980

n.a.: not available

the internationalization of patenting is generalized, significant cross-country differences emerge. In 1988, the ratio was particularly high for R&D intensive small and medium sized countries, such as the Netherlands (7.9), Switzerland (7.6), Belgium (7.0), Sweden (6.5), and Denmark (6.4). Countries which have a small internal market are to some extent forced to appropriate the results of their innovations in foreign markets. Larger countries, on the contrary, have a weaker propensity to extend their patented inventions abroad.

The US and the UK have a lower ratio than the other advanced countries. The very low value of this index for Japan should be noted: for any three domestic applications, there is only one application abroad. This result is essentially an institutional artifact, since the Japanese patent system does not allow the inclusion of the different technical aspects of the same invention in one application, thus inducing inventors to multiply their applications in the domestic market. On the other hand, however, this result also indicates that, in spite of the fast growth of Japanese patents abroad already experienced over the last decade, this country still has vast patent potential.

3.4. Evidence from Bibliometric Indicators

Scientific activity is one of the sources of innovation and contributes to the advancement of technological knowledge. activity in the advanced countries, an analysis of the indicators of scientific efforts may shed new light on the characteristics of the national systems of innovation and on the strategies pursued by individual countries, while offering a comparative perspective on the processes of specialization in science and technology.

We are aware that the distinction between "scientific" and "technological" activities is extremely loose, in spite of the theoretical and empirical efforts made to discriminate between them.[13] In this report, we will distinguish science and technology on an institutional basis, using the term "scientific" for activities carried out in institutions committed to the advancement of knowledge and producing non proprietary results, in contrast with activities labelled "technological", which are intended to generate new or better products and processes, bound to competitive assessment and proprietary in nature.

In the section above "technological" activities were considered by using patenting indicators, while in this section "scientific" activities will be considered by using bibliometric indicators. Publication of research results is one of the main aims of scientific activity, but teaching and other activities of applied research, which do not lead to scientific publications, are also important parts of the scientific enterprise.

In order to describe and investigate the characteristics of scientific activity, a variety of bibliometric indicators have been developed, aimed at measuring the quantifiable aspects of scientific literature. Articles in scientific journals, books, conference proceedings and technical reports offer important information on the amount, object and content of research activities. These indicators, developed from a growing number of databases, are used as output indicators of scientific activity in countries, sectors and institutions. There is a large literature describing the nature of scientific publications and citations.[14] The main indicators of scientific activity are the number of papers, and the number of citations they receive in other papers. In particular:

- the count of scientific papers provides the most general output indicator of scientific activity. Simple aggregations of data by country and by scientific field or subfield allow an overview of the structure of scientific activity and of its changes over time.
- the count of citations provides information on the impact of scientific papers. Citation data (for countries, institutions and authors) can be ordered in two ways: a) they can refer to citations received from other papers appeared after the cited publication, thus showing the impact a paper (or a set of papers) has had on the scientific literature; b) they can refer to citations made to previous publications, showing the influence of past research on current literature. The former approach is the one that will be used here.

While the number of citations a paper has received is a useful indicator of its impact on the scientific literature, it is not an adequate measure of the quality of scientific work. Several factors, including the constraints of dominant scientific paradigms, the underestimation of applied studies and the lack of content evaluation, limit the reliability of this indicator, which tends to reduce the complexity of scientific research (see Lindsey, 1989). Particular caution is therefore needed in the evaluation of citation data.

International scientific activity is so extensive and diverse that no database can include all the publications. The available bibliometric databases consider a subset of world publications selected in a variety of possible ways. The data base used in this report has been developed by CHI Research for the US National Science Foundation, using data from the Science Citation Index of the Institute for Scientific Information. Articles, notes and reviews published by a number of important international journals are considered. The first database includes bibliometric data from 1973 to 1984 drawn from a constant set of 2100 scientific and technical journals, which were selected in 1973. The second database includes data from 1981 on, based on a set of 3081 journals chosen in 1981.[15]

Annual data for all countries on the number of publications, the number of citations received from later papers, and the citations to previous papers are available in both databases; all these indicators refer to publications appearing in the fixed journal set. The same database will be analysed by scientific field and subfield in Chapter 7.

It should be stressed that this database, however large, represents only a subset of the world's scientific literature (and the same applies to any other existing database). The database is dominated by English-language journals and biomedical sciences are highly represented because it was originally developed to provide up-to-date information on publications in these areas. However, analysis across countries and over time is not affected too much by these characteristics of the database; as English has increasingly become the international language of the scientific community, we can expect that most internationally relevant contributions to scientific advance would appear (also) in English.

In order to avoid annual fluctuations in the data, and to ensure a substantial number of cases even in the most disaggregated sectoral distribution, the 1973 journal set annual data series has been grouped into two periods, 1973–78 and 1979–84. The data of the 1981 journal set have been grouped in a single period 1981–86 for the same reasons. An analysis of the changes over time in the patterns of scientific activity is therefore possible for the first database. Three sets of data for number of papers and number of citations are used in this analysis:

1) 1973–78 data based on the 1973 journal set (A1 data);
2) 1979–84 data based on the 1973 journal set (A2 data);
3) 1981–86 data based on the 1981 journal set (B data).

A preliminary analysis of the countries' positions in aggregate terms is developed here, while Chapter 7 will examine the patterns of sectoral specialization of scientific activity in advanced countries.

The percentage distribution[16] among countries of world papers and paper citations in the three databases is shown in Table 3.8.[17] These data provide an overview of the relative importance and pattern of change of the countries considered; national shares, however, reflect the characteristics of the database, where – as it was already pointed out – English language journals account for a disproportionate share of all publications.

The US accounts for 37.1% of all papers and 52.6% of the citations included in the database in both the 1973–79 and the 1979–84 periods. In the second database, the US share falls slightly to 35.8 and 50.6%. EEC countries show a remarkably stable pattern, with about 26% of all papers and 25% of citations in all periods; the largest contribution comes from the UK, whose share falls from around 9.1% in 1973–78 to 8.3% in 1981–86;[18] Germany and France show a slight fall in the number of papers and an increase in their share of citations; Italy, starting from a very low level, increases in both respects. Japan, on the other hand, expands its share of world papers from 5.6% in 1973–78 to 7.3% in 1981–86, and its citations rise from 4.2% to 5.8%. The remaining OECD countries display a generally stable pattern.

Articles published in English are likely to be cited more often than those published in other languages since English is the most widely used language of the scientific community. In fact, the US has the highest number of citations per paper, on average, in all periods, but is followed closely by Switzerland (especially in the latest data), Sweden, Denmark, the UK, and the Netherlands. The EEC aggregate is further behind, but over time moves closer to the world average; a similar convergence can be found also for Japan, which starts from an even lower average number of citations per paper. Combined with the other science and technology indicators used in this study, such data provide an additional insight into the structure of research activities in the advanced countries; the sectoral analysis of scientific activities in Chapter 7 will allow more specific links to be made between science and technology.

3.5. Comparison of Countries in Terms of Different S&T Indicators

The indicators used in this Chapter provide information on different aspects and phases of the S&T process, which also reflect the nature of national innovation

TABLE 3.8

SCIENTIFIC PAPERS AND CITATIONS OF ADVANCED COUNTRIES

Percent distribution across countries of the number of papers and paper citations 1973-78, 1979-84, 1981-86

Countries	1973-78 DATA PAPER	CITAT.	1979-84 DATA PAPER	CITAT.	1981-86 DATA PAPER	CITAT.
United States	37.17%	52.62%	37.05%	51.60%	35.85%	50.59%
Japan	5.60%	4.21%	7.14%	5.70%	7.26%	5.85%
EEC 12	26.54%	24.72%	26.00%	24.76%	26.32%	25.57%
W. Germany	6.42%	5.22%	6.18%	5.85%	6.07%	5.84%
France	5.49%	3.92%	5.14%	4.27%	4.84%	4.31%
Un. Kingdom	9.10%	10.67%	8.26%	8.99%	8.30%	9.24%
Italy	1.82%	1.27%	2.08%	1.57%	2.31%	1.75%
Netherlands	1.35%	1.57%	1.55%	1.75%	1.72%	1.92%
Belgium	0.82%	0.73%	0.81%	0.78%	0.87%	0.85%
Denmark	0.80%	0.99%	0.86%	0.97%	0.84%	0.92%
Spain	0.38%	0.18%	0.68%	0.36%	0.89%	0.47%
Ireland	0.19%	0.09%	0.18%	0.10%	0.17%	0.11%
Portugal	0.04%	0.02%	0.05%	0.03%	0.06%	0.03%
Greece	0.14%	0.06%	0.21%	0.09%	0.24%	0.11%
Switzerland	1.38%	1.57%	1.34%	1.76%	1.29%	1.78%
Sweden	1.58%	2.17%	1.65%	1.93%	1.68%	1.99%
Austria	0.58%	0.29%	0.55%	0.34%	0.55%	0.35%
Canada	4.30%	4.31%	4.20%	4.05%	4.14%	4.00%
Australia	1.76%	0.63%	2.01%	1.87%	2.14%	2.02%
Others	21.10%	9.50%	20.07%	7.98%	20.76%	7.87%
WORLD	100%	100%	100%	100%	100%	100%

Source: CNR-ISRDS elaboration on CHI Research data

systems. The country-specific elements in S&T activities include the following:

i) The relative role of the business and public sectors in financing and performing R&D (see Table 3.5).

ii) The interaction between the business and the public sectors. In some countries, universities and public research centres promote active collaboration with the business sector; in other countries vast government research programmes are contracted to firms; yet in other countries little cooperation would be found between the academic and the business world.

iii) Nations also differ in their willingness and ability to undertake inventive and innovative activities, in their assessment of the risk involved and in their time horizons. Some countries devote more effort to "basic" research as opposed to countries focusing on "applied" and "development" research.

As already pointed out, the indicators used here contain a certain amount of distortion. For example, bibliometric data are biased by the large number of English language journals included in the data set. In the case of patents, the number of patents registered in any of the institutions considered is biased by the number of applications presented by domestic inventors. Therefore, the share of patents in each institution is not an accurate and comparable measure of the different countries' technological capabilities. Comparison between two countries can however be made by using the patents registered in a third, and equally important, market. This comparison shows, for example, that the number of patents held by Japan alone in the US is approaching that of the twelve European Community member countries. The share of patents held by the US in Japan is slightly above the EEC, while the Japanese patents at the EPO amount to 60% of US patents (see Table 3.6). These data on patents present a picture of the relative strengths of the three main industrial areas which partially differs from the results obtained for R&D expenditure. Japan has a patent share higher than its R&D expenditure compared to both the US and the EEC. The US share of patents is, on the contrary, lower than its share of R&D in comparison to both Japan and the EEC. Consequently, the EEC has a patent share which is, in comparison to its R&D, higher than the US, but lower than Japan.

A similar comparison can be made by matching R&D and patent data with the evidence from scientific publications. Figs. 3.1, 3.2 and 3.3 compare the position of individual countries with regard to these three indicators, which have been trasformed into their natural logarithm for ease of presentation. Figure 3.1 shows the number of patents granted in the US (1982–88) versus R&D expenditure (1982–88), Fig. 3.2 the number of scientific papers (1981–86) versus R&D expenditure (1982–88), and Fig. 3.3 the number of patents granted in the US (1982–88) versus the number of scientific papers (1981–86). Obviously, the three measures are strongly and positively correlated, as shown by the regression lines indicated in the figures in order to illustrate the overall pattern.[19]

Figure 3.1 shows that Japan, West Germany, Sweden and Switzerland display comparatively better performance in patents than in R&D. In all these countries, the R&D financed and performed in the business sector is higher than elsewhere. France, Italy and Spain, which are characterized by a lower contribution of the business sector to the total R&D expenditure, have an opposite pattern, with a poorer performance in patents with respect to R&D. Figure 3.2 shows that both the US and the EEC are characterized by an above average number of papers, in comparison to their R&D expenditure. Japan, on the contrary, shows a lower effort to publish in scientific journals in comparison to its R&D. Within Europe, the UK has relatively more papers than R&D expenditure, while Germany, France and Italy show the opposite trend. Finally, Fig. 3.3 shows the position of each country in the two output indicators of patenting and bibliometrics. The EEC aggregate has a relatively higher output in scientific than technological activities, just the opposite of Japan. Within the EEC, however, Germany has a comparatively larger amount of patents than publications, while the UK, and to a lesser extent France, place greater emphasis on scientific activities. Again, these differences seem to reflect the emphasis of national innovation systems towards activities aimed at scientific or technological results.

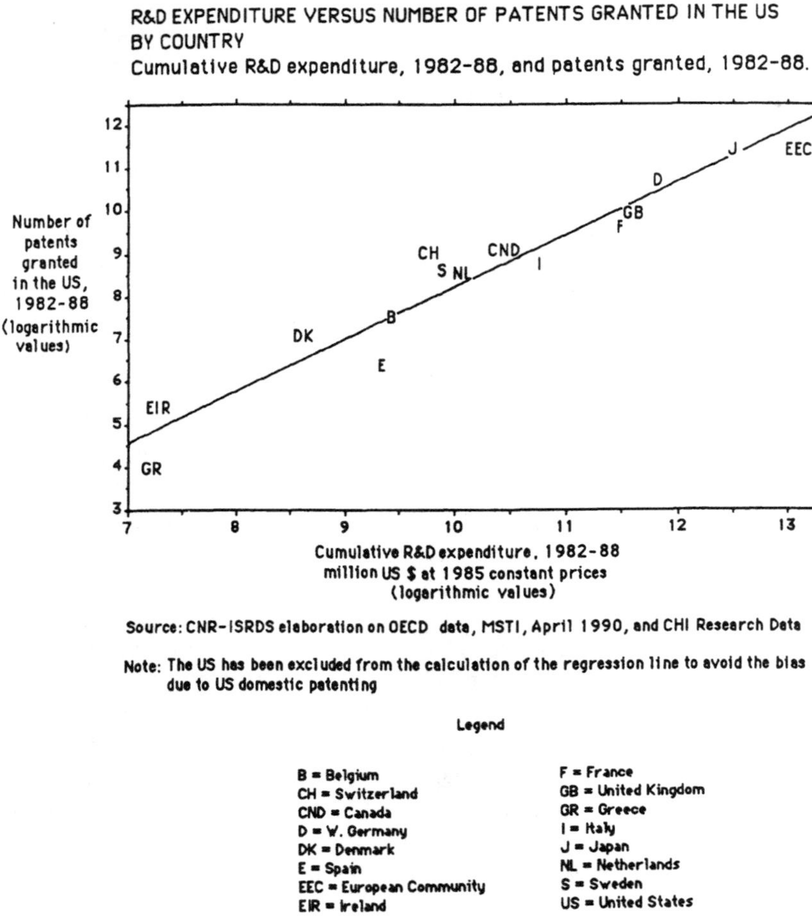

Fig. 3.1.

3.6. The Convergence of S&T Patterns among Advanced Countries

Althought highly aggregated, the data presented in this Chapter show a few significant features of the current innovative process. The first significant result is the considerable increase over the past decade in resources devoted to R&D in all countries; they have generally grown at a higher rate than GDP. Signs of a slowdown have however emerged since the mid 1980s in several countries. This result may identify a "limit" to the increasing role of formalized knowledge in production which characterizes modern economies. Technology developed outside R&D laboratories might have become more important, acting as a substitute for knowledge developed inside them. The generation and use of knowledge has to be coupled in organizations in order to introduce innovations successfully. These factors also seem to be linked to the growing number of channels for the acquisition and transfer of technology, ranging from firms' networking, inter-firm technical agreements, takeovers of other firms, institutional cooperation between the busi-

R&D EXPENDITURE VERSUS NUMBER OF PAPERS BY COUNTRY
Cumulative R&D expenditure, 1982-88 and number of papers, 1981-86

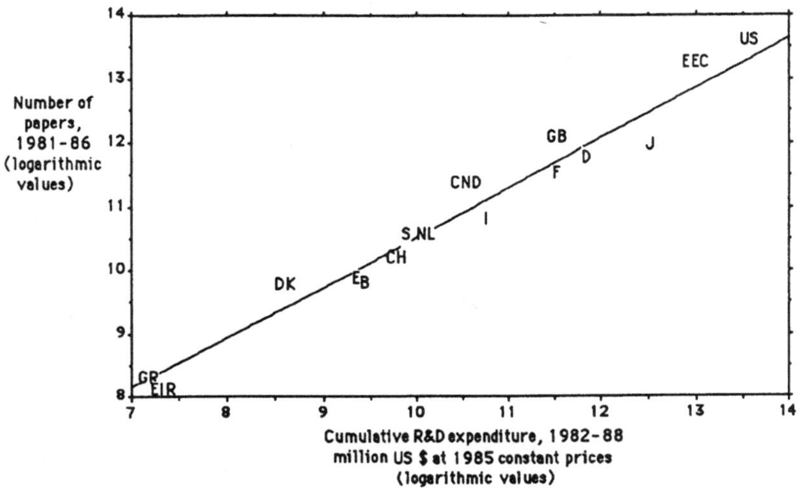

Source: CNR-ISRDS, elaboration on OECD data, MSTI, April 1990, and CHI Research Data

Legend

B = Belgium	F = France
CH = Switzerland	GB = United Kingdom
CND = Canada	GR = Greece
D = W. Germany	I = Italy
DK = Denmark	J = Japan
E = Spain	NL = Netherlands
EEC = European Community	S = Sweden
EIR = Ireland	US = United States

Fig. 3.2.

ness and the public sectors. All these new forms enable firms to innovate without developing the entire range of know-how by themselves and make it possible to exploit the economies of scale and of scope.

The pressure to appropriate the efforts of innovation in international markets has also increased strongly over the last decade. Patents registered in external markets have grown dramatically, in spite of the moderate or even decreasing rates of change in domestic patents. Many inventions and innovations are now exploited at the global level, leading to an increasing competition among firms. The slowdown in the growth of R&D activities may also be a result of the increasing global dimension of technology markets: firms can react to the greater competition to appropriate innovations with a more selective choice of their fields of activity, seeking to receive payoffs in several markets. In order to afford the investment needed to carry out innovative programmes, firms need to be able to exploit the results at the international level.

NUMBER OF PATENTS GRANTED IN THE US VERSUS NUMBER OF PAPERS
Cumulative patents, 1982-88 and number of papers 1981-86

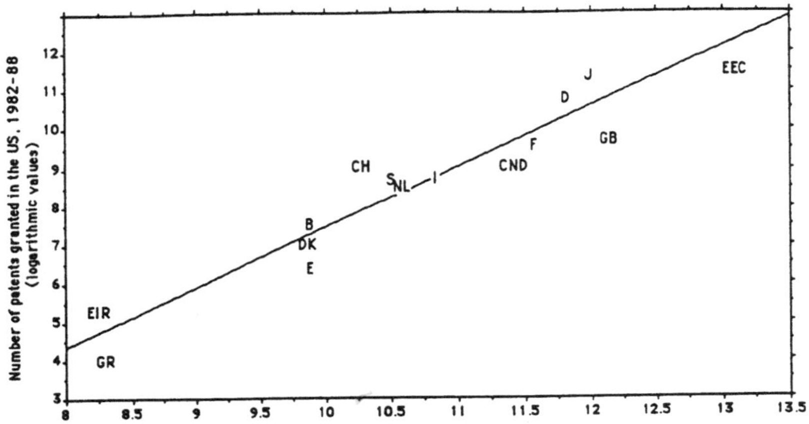

Source: CNR-ISRDS, elaboration on CHI Research Data

Note: The US has been excluded from the calculation of the regression line to avoid the bias due to US domestic patenting

Legend

B = Belgium	F = France
CH = Switzerland	GB = United Kingdom
CND = Canada	GR = Greece
D = W. Germany	I = Italy
DK = Denmark	J = Japan
E = Spain	NL = Netherlands
EEC = European Community	S = Sweden
EIR = Ireland	US = United States

Fig. 3.3.

The second aspect addressed in this Chapter is the position of the main industrial regions. The technological intensity of the US economy is paralleled now by its main competitors, Japan and Germany, but small countries such as Sweden and Switzerland have also joined the group of the most technology-intensive economies. Not all the countries, however, have converged: in spite of their growing effort, the gap between some countries at the periphery, and the core of the most advanced countries has not narrowed. Within the European Community, Germany emerges as the real technological leader, accounting for about one third of total EEC R&D expenditure and for about half of its patents. It should also be noted that our data do not include the former German Democratic Republic, and the German unification in 1990 has further increased its position in S&T. Conversely, there has been substantial erosion of the technological position of the United Kingdom.

The process of European integration seems to have contributed to a greater convergence in S&T activities among the member countries: those starting from quite low levels of technological activity, such as Italy, Spain, Ireland, Portugal, and Greece, have grown at a higher rate than the EEC average. In more general terms, the process of European integration is taking place under two divergent forces. On the one hand we have the centripetal effect towards the core areas, on the other the spillovers towards the less developed regions. While the former force has prevailed over the latter in the process of economic integration, the opposite trend has hitherto prevailed in the field of S&T activities. In spite of the tendency towards EEC convergence, disparities in S&T are still far greater than for economic variables such as GDP per head.

It is, however, difficult to predict whether the process of convergence will continue in the future with the single market after 1992. Europe is experiencing fundamental changes which will strongly affect its S&T system. Different options are open. In one scenario, the process of integration might lead to an increasing division of labour, with only a few countries generating S&T knowledge and diffusing it to all member countries. An alternative scenario might lead to all countries reaching similar S&T intensity, but this would necessarily imply a greater selectivity in sectoral specialization. Aggregate data cannot tell us much beyond that. The sectoral examination developed in the following Chapters will provide detailed evidence to address this question.

Notes

1. National data have been converted in US dollars by using the OECD Purchasing Power Parities provided by the Main Science and Technology Indicators statistics.
2. Japanese R&D national sources tend to overestimate national expenditure in comparison to the other OECD countries. In this report, the OECD adjusted values for Japan have been used.
3. Throughout this book, data will always refer to West Germany.
4. See, among a large literature, Kaldor (1981), Melman (1983), Nelson (1984), Rosenberg (1982, 1986).
5. For surveys on the use of patenting as a technological indicator, see Basberg (1987), Griliches et al. (1987), Pavitt (1988), Griliches (1990).
6. On the effectiveness of the various methods of appropriability, see Wyatt et al. (1985), Levin et al. (1987). For a survey, see Archibugi and Malaman (1991).
7. For a discussion of this indicator, see Narin and Olivastro (1988b), and Trajtenberg (1990).
8. While the countries' shares of R&D expenditure reported in the previous section refer to the OECD area only, the countries' share of patents refer to all countries. The share of patents held by non-OECD countries is, however, modest.
9. In this report, we will consider patents *granted* in the US, and patent *applications* to the EPO. The latter institution started in 1979 only, and since there is an average time lag of 2–3 years to examine an application, a significant time series for patents granted is not yet available.
10. In the US, almost two thirds of the applications are granted a patent, and at the EPO more than 80%.
11. A single invention is counted as many times as the patent is extended in different countries.
12. Cf. Schankerman and Pakes (1986), p. 1071: "Part of the decline in patenting per unit of inventive input may reflect a shift away from "more patents" to "higher quality"".
13. Cf. Rosenberg (1982), Price (1984), Archibugi (1986). An empirical study of this relationship is found in Narin and Noma (1985).
14. See among others, Small and Griffith (1974), Griffith et al. (1974), Garfield (1979), Lindsey (1989). For an overview of science indicators, see Van Raan (1988).
15. The number of journals considered in both databases falls over time as some journals cease publication. The second database has been developed in order to include several new journals and to extend the coverage of the database. The papers appearing in a journal are assigned to the field and subfield of science to which the journal is related.

16. As noted above for patenting, the countries' shares considered here refer to the publications of all countries. Non-OECD countries hold a share of publications considerably larger than their share of patents.
17. As noted above with regard to patent citations, a warning should be considered when the number of citations is calculated by grouping data for several years, since citations refer to papers of different ages and therefore with a different probability of being cited in subsequent papers.
18. The relative decline of British scientific output has been the subject of a long standing discussion, which has also raised questions as to the reliability of the databse used here to assess countries' positions (see Martin et al. (1987), Leydersdorff (1988), Braun et al. (1989), Irvine and Martin (1989)). For our purposes of broad cross-country comparison, however, the CHI Research database can be regarded as valid.
19. The EEC aggregate has not been included when calculating the regression. The US is excluded in the two figures presenting data on the US patent system. As the figures are on logarithmic scale, the relationship between variables is not linear.

CHAPTER 4

Sectoral Strengths and Weaknesses of Advanced Countries in Different Patent Institutions

4.1. The Analysis of Sectoral Specialization

The assessment of national capabilities and performance in specific fields of technology is becoming increasingly important for the policy making of both governments and firms. The previous Chapter shows that in all countries the technological effort, measured both by input indicators such as R&D and by output indicators such as patents, has grown over the last decade. A tendency towards an increasing internationalization of technological activities has also been identified. Moving beyond the aggregate picture of science and technology (S&T) activities, we examine in this and in the following Chapters the sectoral patterns of national specialization in technology, using patent data as an indicator. This evidence will then be matched in Chapter 7 with a parallel analysis on the sectoral patterns of scientific specialization, using bibliometric indicators.

The analysis of these Chapters makes possible to identify the fields in which a country has an advantage or a weakness relative to its overall scientific and technological activities. The technological specialization of countries in particular sectors is the result of several factors. Firstly, the cumulative nature of know-how in production and technology means that a country's economic history will play a key role in shaping its current patterns of specialization. Secondly, a variety of factors rooted in the structure of a country's economy affects its comparative advantages in technology. These include: i) the sectoral specialization of production and trade shown by the country's industries; ii) the existence of natural resources and of a domestic industry based on their exploitation; iii) a particular national demand structure and consumer tastes which may lead to specific technological developments. Thirdly, government policies concentrating national resources in particular technological fields can result in a country's relative specialization in these areas. The most obvious examples include the state-supported and publicly-funded activities of military industry, nuclear technology, and space programmes.

Some of the above elements will be clearly identifiable in interpreting the findings of the analysis of technological specialization in this and in the following Chapter. Still, the pattern of sectoral strengths and weaknesses in technology can hardly be predicted on the basis of a country's industrial structure or government policies alone. Although technological accumulation and economic activities are crucial in shaping the national profiles of specialization, a country's technology has a dynamics of its own; countries starting from the same knowledge can evolve along different trajectories and similar industrial structures can be associated to widely differing technological intensities. From this approach to the problem of a country's technological specialization, it follows that a substantial degree of detail is needed in empirical analysis in order to identify the specific factors resulting in relative strengths and weaknesses in individual fields. Technological knowledge is, in fact, highly diversified not only across countries but also across industries and firms, where different innovative strategies are followed and specific technology policies are required.

A set of methodological problems has to be addressed in approaching a disaggregated analysis of the sectoral distribution of the technological activities of advanced countries, exploring their structure, similarities, differences and changes over time. In order to make sure that the relevant patterns emerge from the empirical investigation, the type of sectoral disaggregation available for the data under examination must first be considered. Methodological and statistical problems are relevant in selecting the appropriate degree of disaggregation; if the sectors of analysis are too broadly defined they may conceal the specialization processes occurring within them; on the other hand too detailed a disaggregation, based on classes with widely different size and importance, complicates the analysis and makes interpretation of the results more difficult.

An additional point regards the logic of the classification adopted. For example this Chapter uses patent data classified according to International Patent Classes (IPC) and generally based on human necessities or technical functions, while patent data disaggregated according to the Standard Industrial Classification (SIC) and based on product groups will be used in the next Chapter. Within a given classification of science and technology activities in different sectors, a second point in the study of specialization is the problem of developing statistical methodologies which make it possible to compare the relative position of each country across fields. Appropriate indicators can make the specialization indexes independent of the countries' size and specific fields of activity, allowing a comparison across countries of their relative advantages and weaknesses. The statistics used will be described in section 4.3.

A third aspect of the analysis of specialization regards the changes of previous statistics over time, which can show whether a country has increased its strengths in selected areas or shifted its relative advantage to new fields. Fourthly, an overall assessment of the degree of specialization is possible, showing to what extent national activities are concentrated in a few areas or spread across most fields; again, this makes it possible to carry out comparison both across countries and over time. The interest of investigating specialization patterns in science and technology is not limited to the description of the fields of activity of individual countries; it can be related to the processes of specialization in industrial production and international trade, contributing to the explanation of national performances and innovative activities, as will become clear in the following Chapters.

4.2. Studies on Sectoral Patenting Activities

Patent data are a well-known indicator of technological activity making international comparisons and sectoral studies possible; a large literature has pointed out its limits as well as its significance (see Soete and Wyatt (1983), Basberg (1987), Pavitt (1988), Griliches (1990)). Patent data are available for most countries on a standard basis. The World Industrial Property Organization (WIPO) publishes an annual report containing the number of patent applications and patents granted in each country, broken down by country of origin and International Patent Classes (IPC) at the two digit level.

Within the Commission of European Communities, Grevink and Kronz (1985, 1986) provided data on patent applications in selected EEC countries and in the US, broken down according to the three digit IPC and NACE classifications. Several national patent offices regularly publish data on patent applications received and patents granted (see, among others, the European Patent Office annual report).

The use of patenting as an internationally comparable indicator at the sectoral level was pioneered by Luc Soete (1981, 1987). Considering patents granted in the US, Soete seeks to identify the role of technological innovation in international competitiveness. A concordance between the Standard International Trade Classes (SITC) and patent classes has been produced in order to identify the technological determinants of international competitiveness. Soete assumes that industrial countries have a similar propensity to protect their inventive activity in the US, but the same assumption is not made for US firms, since firms of every country have a higher propensity than their foreign competitors to patent in their home market. The sectoral technological potential of the US economy is therefore estimated on the basis of R&D expenditure.

The Science Policy Research Unit of Sussex University has undertaken a vast program of international comparisons based on patent statistics. More particularly, a data base of patents granted in the US has been created which contains information for countries, firms and technological classes (at the two, three, and four digit levels). Patel and Pavitt (1987a) compare the sectoral specialization of the major industrial areas, i.e. Western Europe, the US and Japan. The data used, however, do not allow immediate comparison between the US and other countries, because of the bias in favour of the US firms and inventors pointed out above. Patel – Pavitt (1987b), and Pavitt – Patel (1990) have elaborated technological profiles for the United Kingdom and France based on patents in the US and have been able to relate the technological strengths and weaknesses of these countries to the patenting activity of some major companies. Some international comparisons between levels of sectoral concentration and firms' behavior have also been performed for the United Kingdom and West Germany (Patel – Pavitt, 1989b).

A full data base on US patents and related patent citations is available from CHI Research (for a full description, see Narin – Olivastro, 1988b). This database contains information on countries, firms and technological classes (at the two, three and four digit levels) for both patent and citation counts. Such data will be used in Chapter 5 to make further international comparisons. By means of this source, Narin and Olivastro (1987a, 1987b, 1988a) have identified areas of excellence for the United Kingdom, West Germany and Japan. On the base of the same source, a patent profile of, and a comment on, the top 50 Japanese firms has also been produced (Chi Research, 1988). Additional studies of sectoral patenting activity have been developed by German researchers (Grupp and Legler (1989), Grupp and Schwitalla (1989)), and by Engelsman and Van Raan (1990) in the Netherlands. The latter have examined the methodological problems of patent indicators and compared the Dutch position in technological fields with six major countries.

We have studied the sectoral structure of innovative activities in Italy using extensively patenting indicators (Archibugi (1988a, 1989), Pianta and Archibugi (1991)). It was shown that the Italian sectoral specialization measured by patents extended abroad is fairly consistent in the different countries where patents are registered. Other studies on patenting activities in Italy include those by CER (1988), Sassu and Paci (1990) and Paci (1990), which compare the sectoral distribution of Italian patents registered in different countries, confirming the differences between the areas of activity in domestic and foreign markets. Historical patterns of technological accumulation have been identified by Cantwell (1989) using patents extended in the US. Cantwell and his collegues at the Department of Economics of Reading University have developed a database on the patenting activity of multinational firms using their patenting in the US.

The majority of the studies here reviewed have used patents granted in the US (for significant exceptions, see Schmoch et al. (1988), Schmoch and Grupp (1989)). This comes as no surprise: if a single country has to be selected as a basis for international comparison, it is understandable that the US should be chosen. The US remains the most technologically advanced and the largest single market in the world. It is therefore reasonable to assume that firms of all industrial countries have an interest in protecting their most significant inventive and innovative activities with patents in the US. Some case studies (cf. Basberg (1983), Archibugi (1988a, 1989), Sassu and Paci (1990)) have shown that this is so. Moreover, data on patents granted in the US have been computerized for a long time, which has unquestionably provided an additional incentive for scholars to use them.

However, the assumption that patents extended in the US are a reliable measure of the technological specialization of advanced countries has never been tested adequately. Moreover, US patents cannot tell us anything decisive about the technological specialization of the US because of the substantial difference between the patent strategy pursued by firms in domestic and foreign markets. For these reasons, we have developed a database including patent data drawn not only from the US, but also from other major patenting institutions, such as the European Patent Office, France and West Germany.

4.3. The ISRDS Database and the Analysis of Data

A key issue addressed in this Chapter is the investigation of the consistency of patent data broken down by technological sector in different patent institutions. In order to overcome the limitations of previous studies which rely on patents in the US alone, this section introduces a new set of data developed in our Institute and referred to as ISRDS database. Patents registered at the European Patent Office (EPO), France

TABLE 4.1

Percentage distribution by technological field
International Patent Classes

IPC	Patents grant. in the US to non residents	Total patents in the US	Total patents at the EPO
1 Agriculture	1.01%	1.44%	1.24%
2 Foodstuffs	1.05%	1.15%	1.31%
3 Footw.clothing	2.02%	2.83%	2.16%
4 Health	3.17%	4.71%	3.93%
5 Medical	2.53%	2.22%	2.06%
6 Separating	4.61%	4.92%	3.45%
7 Machin.tools	3.67%	3.33%	2.75%
8 Hand tools	4.61%	4.80%	3.67%
9 Printing	1.41%	1.24%	1.65%
10 Transport	4.30%	4.01%	3.88%
11 Machinery	4.26%	4.78%	3.83%
12 Inor.chemic.	2.04%	2.02%	2.60%
13 Org.chemic.	5.76%	5.34%	9.31%
14 Org.compounds	3.52%	3.68%	4.72%
15 Paint,petrol.	1.87%	2.15%	3.10%
16 Bio-chemistry	0.76%	0.72%	2.00%
17 Metallurgy	2.29%	2.06%	2.33%
18 Textiles	2.29%	1.56%	1.78%
19 Paper	0.42%	0.37%	0.49%
20 Building	2.31%	2.75%	2.95%
21 Mining	0.65%	1.23%	0.61%
22 Engines	4.90%	3.84%	2.62%
23 Engineering	4.47%	4.30%	3.69%
24 Light.&heat.	2.85%	3.23%	2.26%
25 Weapons	0.68%	0.87%	0.66%
26 Optics photo	11.49%	9.90%	8.98%
27 Computing	3.52%	3.65%	3.94%
28 Inform.instr.	3.05%	2.57%	2.94%
29 Nuclear physics	0.47%	0.44%	0.61%
30 Electricity	8.77%	8.96%	9.16%
31 Electro.telecom.	5.07%	4.74%	4.68%
32 Others	0.19%	0.22%	0.62%
All classes	100.00%	100.00%	100.00%

Source: CNR-ISRDS elaboration on WIPO and EPO data

and West Germany are considered, as well as those registered in the US. Although already available at the sectoral level from WIPO and EPO sources, these data have not been systematically used in a study of the technological specialization of advanced countries.

In order to draw comparisons between the patent profiles in different countries, we are obliged to use International Patent Classes (IPC) at the two digit level, which is the classification generally used by patent institutions. This classification, however, does not entirely satisfy our research needs for the following reasons.

i) It is based on human needs or technical functions instead of technological criteria, grouping together for instance all activities related to health, whatever the industrial group they may belong to (ranging from chemicals to services, etc.);[1]

ii) The economic activities related to an IPC class can therefore be difficult to identify. We have tried to solve the latter problem by re-naming some classes in a more significant way from an economic viewpoint. In the next Chapter we will be able to use, for patents granted in the US only, the SIC classification based on product groups and industrial and technological criteria.

In summary, the data base developed at ISRDS on patents classified according to IPC classes includes the following:

a) patents granted in the US from 1981 to 1987;

b) patent applications to the EPO from 1982 to 1987;

c) patent applications and patents granted in West Germany from 1982 to 1987;

d) patent applications and patents granted in France from 1981 to 1987.

The countries considered in this Chapter are: the United States, Japan, West Germany, France, the United Kingdom, Italy, the Netherlands, Belgium, Switzerland, Sweden and Canada. For Canada, however, we have not been able to consider patent applications at the EPO. The time period considered includes the most recent years for which data are both available and reliable; as sectoral patent data show significant fluctuations on a yearly basis (see, for example, the findings of Engelsman and Van Raan (1990, p. 24)), we have summed together all the years considered, in order to have a large number of cases in all classes and thus obtain a reliable distribution across sectors.

An overview of the sectoral distribution of total patents in the US and EPO is presented in Table 4.1, showing the percentage distribution of patents granted in the US to non residents, total patents in the US, and patents at the EPO. The marked differences in the shares of patents are the result of the way classes are defined and of the varying interest in the use of patenting as a tool for protecting innovations in different sectors. In fact, two sectors using similar amounts of innovative resources can show great differences in their patenting activities. In the fields of chemicals, for instance, patent protection is systematically used, while in the field of nuclear physics patenting is less frequently used to appropriate returns from innovative activities. Patents in different sectors therefore incorporate a different amount of resources devoted to S&T and, following Scherer (1983), we will define the sectoral propensity to patent as the ratio of the number of patents in each sector to the total amount of resources devoted to S&T in the sector.

The data presented in Table 4.1 do not, however, take into account the research inputs used, and allow only a comparison of the numerosity of individual classes. Classes with larger shares of total patents are fields where either the propensity to patent is strong, or the size of particular markets is important, or where a large number of innovations actually occur. This evidence indicates that the absolute number of patented inventions is not a significant and reliable indicator of the technological dynamism of a specific class. In fact, it makes little sense to consider a patent in the class of Computing on the same footing as one in the class of Health and amusements, as their economic and technological relevance differ widely. The differences in sectoral propensities to patent are fairly stable across countries and the comparison in Table 4.1 of the shares of IPC classes in patents in the US and at the EPO highlights the different structures of the US and European markets for technology.

In both the US and EPO data, the largest shares are found for patents in Optics and photographic instruments (9–10% of all patents) and Electricity (9%), while the lowest shares are in Paper, Nuclear Physics, Weapons, Mining, Agriculture, Foodstuffs and Printing (all below 1.5%). Within this spectrum of relative importance of individual sectors, a few major market-based characteristics can be identified. The class of Organic chemicals accounts for 9.3% of all EPO patents, but for only 5.3% of US patents; the traditional importance of European organic chemistry, with its German, Swiss, British and Belgian strongholds, is clearly evident here, leading to a greater share of patents at the EPO than in the US in a class where a high propensity to patent is also found. A larger share of patents at the European Patent Office than in the US is also found in Organic compounds, Bio-chemistry, Paint and petroleum and Electricity. The share at the EPO is significantly lower than in the US for patents in the classes of Optics and photographic instruments, Engineering, Machinery, Hand tools, Separating and mixing, Health and amusements.

A further comparison is possible by examining the first two columns of Table 4.1, listing the percentage distribution of patents granted in the US to non residents, and total patents in the US. This makes it possible to identify any possible sectoral distortion related to domestic patenting by US inventors. The shares of each IPC class differ in the two distributions (and the difference would be even more significant if we compared patents granted to non residents and patents to US residents only). The fields where the share of patents held by non residents is greater than the share of all patents granted in the US include Optics and photographic instruments (11.5%, with a share in all patents of 9.9%), Electronic and telecommunications (5.1% compared to 4.7%), as well as Information and instruments, Engines, Textiles and Machine tools. On the other hand, US inventors tend to concentrate their patents more in General machinery, Health and amusements, Footwear and clothing, Lighting and heating, and in other fields with small shares of patents, generally in traditional technologies (Agriculture, Foodstuffs) or in domestic-based industries (Paint and petroleum, Building, Weapons, etc.).

Two aspects overlap in this evidence. First, the activity of a country's inventors tends to be greater when measured in their "domestic" market (the US for US inventors, the EPO for European inventors). This explains, for example, the higher shares of EPO patents in chemical-related classes (while the EPO is not a national institution, it is certainly strongly affected by the patenting activities of major European countries). Second, patenting activity tends to be greater in the market where there is more intense competition to reach a dominant position; the high number of foreign (mainly Japanese) patents in the class of Optics in the US reflects the large share of the US market held by foreign producers.

A second problem is how to deal with the variations across sectors in the propensity to use patenting as a tool for protecting invention. An appropriate use of sectoral patent data entails focusing on the relative distribution of a country's inventive activity in each field, compared to its total patents and to the overall distribution of patents in the institution considered. We have therefore used an index casting light upon relative sectoral specialization, which is equal to the country's share of patents in the technological field j to the total share of the patents of the country registered. This index – known as the Technology Revealed Comparative Advantage index or TRCA and first developed in the study of international trade by Balassa (1965) – has often been applied to patenting (see Soete (1987), Soete, Wyatt (1983)), and will be defined in this report as the specialization index:

TABLE 4.2.

FIELDS OF GREATER SPECIALIZATION OF MAJOR OECD COUNTRIES
BASED ON PATENTS GRANTED IN THE USA, 1981-87

Top five IPC classes with the highest index of technological specialization

	Rank of the top five fields of specialization IPC classes and indexes of specialization				
	1	2	3	4	5
USA	Mining 1.37	Health 1.26	Agriculture 1.24	Footw. cloth. 1.22	Weapons 1.17
JAPAN	Inform.instr. 1.90	Optics photo 1.41	Computing 1.41	Electron.tel. 1.41	Engines 1.26
GERMANY	Weapons 1.60	Nuclear phys. 1.55	Paint,petrol. 1.44	Textiles 1.43	Org.chemic. 1.36
UN. KING.	Mining 2.02	Medical 1.88	Paint,petrol. 1.35	Foodstuffs 1.30	Org.chemic. 1.19
FRANCE	Nuclear phys. 3.14	Medical 1.64	Mining 1.48	Weapons 1.40	Engineering 1.36
CANADA	Agriculture 3.58	Building 2.78	Mining 2.41	Footw.cloth. 2.34	Paper 2.26
ITALY	Footw.cloth. 2.44	Medical 2.16	Foodstuffs 2.15	Textiles 2.12	Machinery 1.81
NETHERLAND	Foodstuffs 3.24	Others 2.02	Agriculture 1.97	Electron.tel. 1.97	Inform.instr. 1.58
SWITZERLAND	Textiles 2.82	Org.chemic. 2.42	Paint,petrol. 2.07	Foodstuffs 1.61	Weapons 1.57
SWEDEN	Paper 4.94	Weapons 3.27	Light.& heat. 2.04	Mining 1.90	Building 1.86
BELGIUM	Agriculture 3.31	Paper 2.52	Inorg. chem. 2.46	Weapons 2.06	Paint,Petrol. 1.94

Source: CNR-ISRDS, elaboration on WIPO data

$$(2.1) \qquad \mathrm{TRCA}_{ij} = (n_{ij}/\Sigma_i n_{ij})/(\Sigma_j n_{ij}/\Sigma_i\Sigma_j n_{ij})$$

where n_{ij} is the number of the patents of the country i in the technological class j registered in a specific patent office. This index has the property of being equal to 1 if the country holds the same share in a given technology as in the total of the country's patents, and of being below (above) 1 if there is a relative weakness (strength). An index equal to 2.0 means that the country holds twice as many patents as expected in the class. An index equal to 0.5 means, on the contrary, that the country holds half as many patents as expected.[2]

A value of the index greater than 1 indicates a relative advantage only (i.e. relative to the existing capability of the country), and should not be confused with an absolute advantage. For example, Italy applied for 692 patents in the class *Footwear and clothing* at the EPO from 1982 to 1989, which account for almost 10% of total patent applications in the class. Since Italy's share in this class is three times higher than Italy's share for total patents, the specialization index is rather high, 2.97. The US, on the contrary, has an index equal to only 0.55 i.e. is relatively weak in this class, but the absolute number of its patents is 1001, i.e. in absolute terms its capability in the field is higher than Italy's. This should be borne in mind when interpreting the sectoral results.

In order to avoid a bias in favour of the countries patenting in their home market, the totals used to calculate the indexes in national patent offices (US, France and West Germany) exclude the patents of the host country (with the exception of the EPO since there is no host country for this organization). The indexes for US patents in the US, French patents in France and Germans patents in West Germany are, on the contrary, calculated on the total number (domestic and foreign) of patents. This index makes it possible to rank the areas of specialization and compare them in different patent offices. A summary of the results is provided in Tables 4.2 and 4.3, which indicate for each country the five areas with the highest values of the specialization indices in the US and at the EPO. The data show that in the two major markets, the top areas of a country's specialization are often different, indicating that the ranking of national strengths is sensitive to the markets where patents are registered.

The overall picture is shown in Table 4.4, which presents the indexes calculated on patents granted in the US and on patent applications at the EPO; from now on we will refer to the vector of TRCA indexes for a country as its "profile of specialization" in a given patenting institution. The indexes resulting from patent applications and patents granted in France and Germany are listed in a previous report (Archibugi and Pianta, 1989, which is available on request); in order to show how consistent national specialization profiles are across different patenting institutions, the correlation coefficients of the different TRCA indexes for 31 IPC classes (the residual class of Other patents has been excluded) are shown in Table 4.5. It should be noted that the data drawn from the different patent offices considered here refer to slightly different time periods, because of the different procedures and time lags between the application and the granting of a patent [3].

In the rest of this Chapter we focus on EPO and US data, also because patenting activities in Germany and France have been increasingly affected, as already pointed out in Chapter 3, by the substitution effect related to the growing importance of the European Patent Office, which can provide protection also in these countries. When a country's specialization profile in Germany and France differs substantially from the picture drawn by EPO and US data, this is pointed out in the section below devoted to a short description of national strength and weaknesses.

4.4. Country Highlights

The United States. The US is the country with the largest patenting activity and shows a rather low variability of its specialization indexes in both its internal market and in the EPO, Germany and France. In domestic patenting, the US is relatively strong in *Mining, Health, Agriculture, Footwear and clothing* and *Weapons*, while in foreign markets it proves to have a relative strength in *Mining, Computing, Organic compounds* and *Medical preparations.* Both at home and abroad, the US is weak in *Textiles, Machine tools* and *Transport.* The classes of US strength in

TABLE 4.3.

FIELDS OF GREATER SPECIALIZATION OF MAJOR OECD COUNTRIES
BASED ON PATENT APPLICATIONS AT THE EUROPEAN PATENT OFFICE
1982-87

Top five IPC classes with the highest index of technological specialization

	Rank of the top five fields of specialization IPC classes and indexes of specilaization 1	2	3	4	5
USA	Mining 1.58	Computing 1.52	Org.compounds 1.44	Medical 1.34	Bio-chemistry 1.33
JAPAN	Inform.instr. 2.49	Electron.tel. 1.52	Bio-chemistry 1.40	Printing 1.39	Computing 1.33
GERMANY	Weapons 1.49	Building 1.36	Transport 1.23	Hand tools 1.20	Engines 1.19
UN. KING.	Engineering 1.60	Transport 1.46	Mining 1.34	Building 1.30	Machinery 1.22
FRANCE	Nuclear phys. 2.63	Building 1.65	Light.& heat. 1.60	Transport 1.59	Agriculture 1.54
ITALY	Footw.cloth. 3.05	Textiles 2.99	Others 2.21	Transport 1.84	Machinery 1.69
NETHERLAND	Agriculture 3.34	Foodstuffs 2.64	Paint,petrol. 1.75	Inform.instr. 1.69	Electron.tel. 1.55
SWITZERLAND	Textiles 2.65	Machin.tools 1.80	Foodstuffs 1.78	Footw.cloth. 1.75	Health 1.65
SWEDEN	Paper 4.74	Weapons 3.25	Others 2.54	Machinery 2.25	Health 1.97
BELGIUM	Agriculture 4.33	Textiles 2.37	Mining 2.25	Building 1.85	Metallurgy 1.80

Source: CNR-ISRDS, elaboration on EPO data

both databases tend to fall in areas related to the exploitation of land and natural resources, in health-related fields and in selected high technology areas, either chemical-pharmaceutical, or electronic (such as computing), or military related (weapons).

However, the specific sectoral strengths and weaknesses of the US differ substantially in domestic and in external patenting, with no correlation emerging between data for the US on the one hand and for the EPO, France and Germany on the other. The specialization profiles emerging from different foreign markets are consistent, however, with correlation coefficients always higher than 0.7. This result is particularly relevant since the US technological performance has hitherto been considered on the basis of US registered patents alone (see above). In fact, the US specialization profile measured by patenting is substantially different when taking into account its domestic or its foreign activities.

TABLE 4.4

SPECIALIZATION PROFILES OF ADVANCED COUNTRIES
BASED ON PATENTS REGISTERED IN THE UNITED STATES AND AT THE EUROPEAN PATENT OFFICE

Specialization indexes by technological field
International Patent Classes

N. IPC	United States		Japan		W. Germany		France		United Kingdom		Italy	
	USA Patents Granted 1981-87	Europ. Pat.Off. Applic. 1982-87	USA Patents Granted 1981-87	Europ. Pat.Off. Applic. 1982-87	USA Patents Granted 1981-87	Europ. Pat.Off. Applic. 1982-87	USA Patents Granted 1981-87	Europ. Pat.Off. Applic. 1982-87	USA Patents Granted 1981-87	Europ. Pat.Off. Applic. 1982-87	USA Patents Granted 1981-87	Europ. Pat.Off. Applic. 1982-87
1 Agriculture	1.24	0.66	0.42	0.22	0.66	1.06	1.02	1.54	1.00	1.09	0.84	1.10
2 Foodstuffs	1.07	1.14	0.56	0.60	0.86	0.67	1.02	0.99	1.30	0.82	2.15	1.42
3 Footw.clothing	1.22	0.54	0.61	0.58	0.92	1.13	1.02	1.20	0.92	1.03	2.44	3.05
4 Health	1.26	1.17	0.69	0.52	0.94	0.88	1.03	0.96	1.10	1.01	0.92	0.93
5 Medical	0.89	1.34	0.68	0.97	0.87	0.76	1.64	0.74	1.88	1.12	2.16	0.98
6 Separ. & mix.	1.05	1.06	0.73	0.64	1.22	1.10	0.95	0.85	1.15	1.13	0.85	0.66
7 Machin.tools	0.92	0.67	0.91	0.82	1.09	1.19	0.94	1.00	0.94	1.03	1.09	1.20
8 Hand tools	1.03	0.85	0.85	0.80	1.14	1.20	0.89	0.89	1.02	1.05	1.18	1.33
9 Printing	0.89	1.01	1.25	1.39	1.16	1.14	0.40	0.46	0.60	0.83	1.22	1.50
10 Transport	0.94	0.58	1.02	0.80	1.05	1.23	1.10	1.59	1.06	1.46	0.96	1.84
11 Machinery	1.08	0.74	0.65	0.46	1.14	1.14	0.92	1.06	1.08	1.22	1.81	1.69
12 Inor.chemic.	0.99	1.16	0.81	0.84	1.10	1.04	1.31	1.04	1.05	0.91	0.68	0.66
13 Org.chemic.	0.94	1.03	0.73	1.04	1.36	1.07	1.09	0.73	1.19	1.05	1.53	1.09
14 Org.compounds	1.03	1.44	1.12	1.25	1.34	1.06	0.72	0.43	0.77	0.57	0.97	0.79
15 Paint,petrol.	1.10	1.32	0.69	0.72	1.44	1.15	0.92	0.47	1.35	0.99	0.65	0.43
16 Bio-chemistry	0.95	1.33	1.02	1.40	0.90	0.60	0.84	0.78	0.96	0.83	1.41	0.50
17 Metallurgy	0.91	1.08	1.03	1.22	0.89	0.80	1.03	1.01	0.92	0.81	0.77	0.61
18 Textiles	0.63	0.61	0.76	0.96	1.43	0.99	0.70	0.96	0.72	0.78	2.12	2.99
19 Paper	0.89	1.07	0.41	0.49	1.15	1.10	0.65	0.91	1.01	0.94	1.32	0.76
20 Building	1.13	0.30	0.40	0.26	0.89	1.36	1.18	1.65	1.08	1.30	1.31	1.60
21 Mining	1.37	1.58	0.18	0.14	1.31	0.68	1.48	1.03	2.02	1.34	0.15	0.12
22 Engines	0.78	0.74	1.26	1.20	1.07	1.19	0.77	1.12	0.95	0.91	0.69	1.24
23 Engineering	0.97	0.87	0.78	0.57	1.30	1.08	1.36	1.33	1.13	1.60	1.04	1.12
24 Light.& heat.	1.09	0.65	0.73	0.53	1.01	1.16	1.24	1.60	0.88	1.14	0.95	1.29
25 Weapons	1.17	0.47	0.14	0.22	1.60	1.49	1.40	1.40	1.16	0.99	1.29	0.84
26 Optics photo	0.87	1.15	1.41	1.29	0.77	0.83	0.76	0.96	0.90	1.03	0.58	0.54
27 Computing	1.03	1.52	1.41	1.33	0.64	0.62	0.88	0.80	0.93	0.89	0.62	0.46
28 Inform.instr.	0.85	1.01	1.90	2.49	0.35	0.56	0.53	0.69	0.46	0.51	0.34	0.44
29 Nuclear physics	0.95	1.19	0.60	0.85	1.55	0.87	3.14	2.63	0.80	0.47	0.28	0.15
30 Electricity	1.02	1.07	1.22	1.30	0.87	0.96	1.15	1.20	0.90	0.83	0.57	0.60
31 Electron.telec.	0.95	0.96	1.41	1.52	0.60	0.97	1.17	1.12	0.92	0.99	0.71	0.51
32 Others	1.12	0.77	0.85	0.53	0.97	1.17	1.05	0.65	1.00	0.89	0.61	2.21

Source: CNR-ISRDS elaboration on WIPO and EPO data.

Japan. Japan's pattern of technological specialization is highly consistent across the various patenting institutions, with correlation coefficients always higher than 0.75, and reaching 0.92 for patents in the US and at the EPO. Japan's most significant advantages are found in electrical and electronic areas such as *Information and instruments*, *Electronics and telecommunications*, *Optics & photographic equipment* and *Computing*. The country has its major weaknesses in traditional technologies such as *Foodstuffs*, *Footwear & clothing* and *Textiles*. Japan is also weak in areas

TABLE 4.4 (continued).

SPECIALIZATION PROFILES OF ADVANCED COUNTRIES
BASED ON PATENTS REGISTERED IN THE UNITED STATES AND AT THE EUROPEAN PATENT OFFICE

Specialization indexes by technological field
International Patent Classes

		Netherlands		Belgium		Switzerland		Sweden		Canada
N.	IPC	USA Patents Granted 1981-87	Europ. Pat.Off. Applic. 1982-87	USA Patents Granted 1981-87	Europ. Pat.Off. Applic. 1982-87	USA Patents Granted 1981-87	Europ. Pat.Off. Applic. 1982-87	USA Patents Granted 1981-87	Europ. Pat.Off. Applic. 1982-87	USA Patents Granted 1981-87
1	Agriculture	1.97	3.34	3.31	4.33	1.25	0.58	1.72	1.34	3.58
2	Foodstuffs	3.24	2.64	0.95	0.83	1.61	1.78	0.96	0.86	1.54
3	Footw.clothing	0.68	0.77	1.04	1.62	1.02	1.75	1.71	1.20	2.34
4	Health	0.69	0.62	1.07	0.93	1.01	1.65	1.54	1.97	1.59
5	Medical	0.69	0.67	1.71	0.72	1.37	0.85	0.61	0.83	0.60
6	Separ. & mix.	1.17	1.26	1.30	0.81	0.88	0.93	1.52	1.60	1.10
7	Machin.tools	0.58	0.48	0.88	1.00	1.33	1.80	1.57	1.55	0.83
8	Hand tools	0.96	0.94	1.57	1.31	0.88	0.87	1.49	1.41	1.29
9	Printing	0.45	0.43	0.29	0.27	1.09	0.77	0.69	1.01	0.70
10	Transport	0.49	0.58	0.52	0.86	0.42	0.53	1.21	1.40	1.19
11	Machinery	1.00	1.42	1.26	1.01	1.47	1.60	1.77	2.25	1.45
12	Inor.chemic.	1.15	1.11	2.46	1.41	0.60	0.47	0.78	1.05	1.40
13	Org.chemic.	0.97	0.85	1.11	0.63	2.42	1.58	0.16	0.34	0.39
14	Org.compounds	1.06	0.87	1.55	0.63	0.88	0.52	0.17	0.23	0.47
15	Paint,petrol.	1.29	1.75	1.94	0.62	2.07	1.17	0.32	0.31	0.92
16	Bio-chemistry	0.73	1.09	1.31	1.58	0.80	0.76	0.79	0.61	0.84
17	Metallurgy	0.92	0.53	1.76	1.80	0.94	0.80	0.96	0.72	1.05
18	Textiles	0.59	0.54	1.25	2.37	2.82	2.65	0.59	0.86	0.27
19	Paper	0.48	0.52	2.52	0.66	0.63	0.70	4.94	4.74	2.26
20	Building	1.46	1.41	1.07	1.85	1.09	1.50	1.86	1.55	2.78
21	Mining	1.16	0.77	0.98	2.25	0.26	0.43	1.90	1.54	2.41
22	Engines	0.26	0.44	0.17	0.54	0.46	0.68	0.90	0.93	0.55
23	Engineering	0.71	0.69	0.42	0.67	0.78	0.92	1.44	1.61	0.86
24	Light.& heat.	1.13	1.05	0.84	1.50	0.76	1.06	2.04	1.84	1.58
25	Weapons	0.18	0.27	2.06	1.32	1.57	1.60	3.27	3.25	1.35
26	Optics photo	0.78	0.86	0.95	1.61	0.70	0.88	0.68	0.71	0.66
27	Computing	0.80	0.61	0.38	0.36	1.26	0.96	0.62	0.63	0.76
28	Inform.instr.	1.58	1.69	0.75	0.52	0.25	0.38	0.20	0.30	0.53
29	Nuclear physics	0.47	0.35	1.87	1.25	0.48	0.27	1.83	1.43	0.25
30	Electricity	1.48	1.15	0.52	0.57	0.72	0.58	0.67	0.65	0.79
31	Electron.telec.	1.97	1.55	0.46	0.47	0.49	0.35	0.43	0.29	0.96
32	Others	2.02	0.69	0.31	0.18	0.96	0.72	1.08	2.54	1.25

Source: CNR-ISRDS elaboration on WIPO and EPO data.

where government and military procurement play a key role, such as *Weapons* and *Nuclear physics*. In the chemical areas a less clear specialization emerges: while there is a weakness in classes such as *Health*, *Separating and mixing* and *Paint & petroleum*, in other classes, i.e. *Bio-chemistry* and *Organic compounds*, there are relative strengths.

West Germany. German specialization profiles show marked differences across

TABLE 4.5.

CORRELATIONS AMONG SPECIALIZATION PROFILES OF ADVANCED COUNTRIES IN DIFFERENT PATENTING INSTITUTIONS

Patents in the US, EPO, France and W. Germany
Correlation coefficients between the indexes of technological specialization by IPC classes

	Comparison of patenting institutions										
Countries	US-EPO	US-FRANCE appl.	US-FRANCE grant.	US-GERM. appl.	US-GERM. grant.	EPO-FRANCE appl.	EPO-FRANCE grant.	EPO-GERM. appl.	EPO-GERM. grant.	FRANCE appl.-grant.	GERM. appl.-grant.
United States	0.05 *	0.06 *	-0.02 *	0.26 *	0.18 *	0.82	0.89	0.70	0.85	0.93	0.76
Japan	0.92	0.80	0.76	0.96	0.95	0.89	0.86	0.92	0.91	0.95	0.94
W. Germany	0.49	0.37	0.50 *	0.38 *	0.34 *	0.59	0.59	0.64	0.58	0.93	0.72
France	0.76	0.13 *	0.29 *	0.56	0.82	0.43 *	0.61	0.27 *	0.76	0.93	0.69
United Kingdom	0.54	0.08 *	0.26 *	0.46 *	0.53 *	-0.08 *	0.26 *	0.13 *	0.17 *	0.76	0.74
Italy	0.75	0.82	0.84	0.76	0.78	0.80	0.88	0.69	0.69	0.96	0.79
Netherlands	0.86	0.62	0.66	0.49 *	0.79	0.80	0.82	0.79	0.94	0.98	0.81
Belgium	0.55	0.56	0.68	0.50 *	0.74	0.29 *	0.63	0.46 *	0.46 *	0.68	0.40 *
Switzerland	0.79	0.77	0.80	0.31 *	0.61	0.70	0.89	0.65	0.90	0.82	0.83
Sweden	0.97	0.91	0.92	0.85	0.88	0.91	0.92	0.80	0.88	0.95	0.79
Canada		0.19 *	0.30 *	0.55	0.50 *					0.81	0.56

Source: CNR-ISRDS elaboration on WIPO and EPO data
All correlation coefficients are statistically significant to the level of 0.001%,
except those marked with an asterisk (*)

Correlation coefficients across 31 IPC classes. The residual class "Others" has been excluded

the various patenting istitutions considered. The country has sectoral strengths in *Weapons*, *Building*, *Transport*, *Hand Tools*, *Engines* and in chemical classes at the EPO, and in *Weapons*, *Nuclear physics* and chemical classes in the US. A significant despecialization is found in electrical areas, including *Information and instruments* and *Electronics and telecommunications*. Chemicals are by far the most relevant field of German technological specialization, with a clear strength also in mechanical-engineering classes.

German patents on the home market have also been considered, and we again find a rather low correlation between the specialization profiles based on internal and external patents (the lowest is between patents granted in Germany and in the US: 0.34), but a substiantial difference appears more generally between the areas of greater activity in the US and in European countries. Looking at German domestic patents, the specialization indexes show, as expected, a lower variability than that emerging from patenting abroad.

France. The most significant strengths of the country are found in state supported technological fields, such as *Nuclear physics*, *Medical preparations* and

Weapons for patents in the US and *Building*, *Lighting and heating* and *Transport* for patents at the EPO. It should also be noted that France is one of the few European countries to present a positive specialization in some electrical fields, namely *Electricity* and *Electronics and telecommunications*. The country is, however, weak in other electronic-related technologies, such as *Information and instruments* and *Computing*.

The areas of strength show some difference in the various patenting institutions, and the greatest distance is between specialization profiles based on domestic patents and on patents registered in the US (the correlation coefficient is 0.125); much closer pictures emerge from patenting in the different external institutions. France is the third country for which a comparison between domestic and foreign patenting is possible, and again a much lower variability is found in the specialization indexes for the former, confirming that domestic patents are spread more uniformly across technological sectors.

United Kingdom. A clear pattern of specialization for the United Kingdom does not emerge from its patenting in the different institutions considered. The areas of greater strength include health-related fields, some chemical and mechanical-engineering classes. At the EPO a British advantage is found in *Engineering* and *Transport*; in the US in *Medical preparations* and chemical classes; while *Mining* emerges in both. The country is generally weak in Electrical areas, and more particularly in *Information and instruments*. Areas of weakness are found also in *Textiles*, *Printing* and *Organic compounds*. The correlation coefficients among the specialization profiles are generally low, and are highest between the US and EPO (0.535), and between patenting in France and in Germany.

Italy. Italian technological specialization is fairly uniform across the different patent offices, as shown by the high values of the correlation coefficients. Chapter 3 has shown that the resources devoted to S&T by Italy are modest, and that the ratio between external and domestic patent applications is lower than in other comparable European countries. The country has clear sectoral strengths in traditional industries such as *Footwear and clothing*, *Textiles*, *Foodstuffs,* as well as in *Machinery* and other engineering-related classes. Patents in the US also show a relative advantage in Medical and pharmaceutical-related classes. Italy shares with many other European countries a structural weakness in electronic technologies.

The Netherlands. As we have shown in Chapter 3, the Netherlands is one of the countries with the highest propensity to patent in external markets, in comparison both to domestic patents and to the national resources devoted to R&D. Its patent profile presents two important features. Firstly it is fairly stable across patenting institutions with generally high correlation coefficients. Secondly, it is concentrated in two clusters of sectors: the agriculture-food complex, including classes such as *Agriculture* and *Foodstuffs* and the electrical sectors, in classes such as *Electronics & telecommunications* and *Information and instruments*. Major weaknesses are found in *Transport*, *Health* and *Medical preparations*.

Belgium. Belgium's pattern of specialization is less uniform than in comparable countries, with generally weak correlations across the different patenting institutions. The most relevant sectoral strength found in all of them is in *Agriculture*. The country also presents a specialization in some chemical classes, such as *Inorganic chemicals* and *Bio-chemistry*. On the other hand, Belgium is strongly despecialized in the electrical and electronic classes.

Switzerland. As shown in Chapter 3, Switzerland is one of the countries with the highest patenting activity abroad. Swiss technological specialization is uniform across the different patenting institutions, a result which is consistent with the extensive use of patents made by the country as a means of protecting its innovations. Relevant sectoral advantages are found in *Textiles*, *Organic chemicals*, and *Foodstuffs*. Indexes above one also occur in *Machinery* and *Machine tools*. On the other hand, the country shows remarkable weakness in electronics and allied technologies.

Sweden. Among the countries examined here, Sweden has the highest values of the correlation coefficients across the different patenting istitutions (always higher than 0.75). The country's sectoral strengths are found in *Paper* and in mechanical classes, namely *Weapons*, *Machinery*, *Machine tools* and *Lighting and heating*. On the other hand, Sweden has a weakness in all the electrical related areas, and especially in *Information and instruments*, *Computing* and *Electronics and telecommunications*.

Canada. The US market is the most important destination for the inventions of Canadian firms.[4] Canada has its sectoral strengths in areas connected with raw materials, such as *Agriculture*, *Foodstuffs*, *Paper*, *Building* and *Mining*. On the other hand, the country is comparatively weak in the more science and technology intensive areas, such as the chemical and electronic sectors.

4.5. The Domestic Market Effect on Patenting

The analysis of the database developed at ISRDS on patents registered in the US, the EPO, West Germany and France provides an overview of the sectoral strengths and weaknesses of the advanced countries and new insights into the pattern of technological activities. A first finding is what we have called the "domestic market effect", i.e. the considerable difference between the specialization patterns shown by patents registered in domestic and foreign markets. For the three countries for which we have considered both domestic and foreign patents, i.e. the United States, West Germany and France, very weak correlations are found between the patterns of specialization measured in domestic and foreign patenting. This result is partly due to the presence of individual inventors. Each patent institution receives an above average number of domestic applications in the fields where individual inventors are generally more active, i.e. final products and consumer goods. While individual inventors concentrate their domestic patents in selected technological fields, only occasionally do they extend their patents abroad.

Another factor contributing to the difference between domestic and foreign markets is the lower variability of the specialization indexes in the former than in the latter. A possible explanation of this pattern is that firms tend to protect their inventions (and their domestic market share) also in sectors where they have not developed a world class technological capability, while they tend to patent abroad inventions of greater quality for which there is a higher probability of economic returns. In other words, national firms also patent in the domestic market inventions intended as an additional defence against foreign competitors. These factors can explain the diverging specialization profiles found in domestic and external patenting, and confirm the qualitative differences between these activities, which have already been identified by several case studies (see Savignon et al. (1982), Bertin and Wyatt (1986), Napolitano and Sirilli (1990)).

A second finding of this Chapter is the different degree of consistency of countries' specialization profiles in the patenting institutions compared here. Some countries – including the US, Japan, Italy, the Netherlands, Switzerland and Sweden – show a uniform pattern of specialization across countries (as long as we consider their foreign patenting activities). On the other hand, Britain, Canada, and to a lesser extent France and Germany, are countries presenting relevant differences in their specialization profiles in the various foreign patenting institutions. This can partly be related to the patent data used for the analysis, but is also due to the nature of the sectors of these countries' strengths, which may require a differentiated degree of protection in the various markets examined in this Chapter.

The relationship between the fields of a country's specialization and the specific conditions of a patenting market (including the presence of major competitors and institutional aspects) is an issue which would require more specific study. It is however significant to look at the position of the largest country, the US, for which the specialization pattern on external markets had not been considered in the existing literature. Not surprisingly, the most significant differences are to be found in sectors where innovations can hardly be exploited in European markets, such as Agriculture, Buildings, and Weapons.

A third aspect emerging from this Chapter is the different variability across sectors of the countries' specialization indexes. This finding suggests the need to investigate the degree of national specialization in sectoral technological activities, an issue which will be tackled in Chapter 8.

From this overview of patent indicators and the comparison of patenting activity in different institutions we conclude that important evidence can be drawn from an appropriate use of data on patented inventions in order to analyze the sectoral structure of countries' technological activities. Data based on patenting in large foreign markets do provide a reliable and broadly consistent picture of national innovative efforts in different sectors. On the basis of this evidence, and using patents in the US only, the next Chapter examines a larger number of countries, trends over time, and an indicator of the impact patented inventions have, namely the citations a patent receives from later ones.

Notes

1. To some extent, the IPC classification accounts for vertically integrated sectors.

2. The index is not symmetrical and therefore poses some statistical problems in the analysis of its distribution. In order to describe the dispersion of a country's indexes in the various fields, standard deviation and variance are not satisfactory indicators and in Chapter 8 a new indicator will be used. However, the results found in a previous work in progress using the standard deviations of the specialization indexes of patents in the US, EPO, Germany and France are broadly similar to those obtained in Chapter 8. A variety of statistical refinements of the index have been suggested by Grupp (1989) and Engelsman and Van Raan (1990); they have used the log of the index, and have also explored the measurement of its margins of error.
3. Patents granted in the US in the period 1981–87 may refer to applications of 1 to 3 years earlier (see Schmoch and Grupp, 1989) and therefore there is partial overlap with applications presented at the EPO in the years 1982–87. However, technological specialization tend to be rather stable over time (see Cantwell, 1989). Changes over time in technological specialization will be assessed in the next Chapter.
4. Not surprisingly, the number of Canadian patented inventions is higher in the US than in Canada itself. The number of inventions registered by Canadians in European countries such as France and Germany is much lower than in the US: there are 8,583 patents granted in the US as against 641 in France and 405 in West Germany. Unfortunately, EPO data are not available for Canada.

CHAPTER 5

Changes over Time and Impact of Patenting Activity: The Sectoral Distribution of Patent Counts and Patent Citations in the US

5.1. Assessing the Impact of Patented Inventions

In the previous Chapter we have examined the pattern of specialization of advanced countries across technological fields described by IPC classes, on the basis of patenting in the US, at the EPO, in West Germany and France. The consistency shown by most countries in their profiles of specialization emerging from patents registered in different foreign institutions is documented for the first time and the findings add new depth to our undestanding of the areas of strength and weakness of major countries.

In this Chapter a number of additional issues are addressed: i) the analysis is extended to a larger number of countries, including all 12 EEC countries; ii) changes of specialization profiles over time are investigated; and iii) an indicator of the impact of patented inventions is considered, namely patent citations.

The aggregate number of citations received by the patents registered in the US by each country has already been examined in Chapter 3. Patent citations are an indicator of how important a patented invention is considered by other patents and offer a measure of the impact a patent has on later patented inventions. This does not mean that patent citations reflect the "quality" or the economic value of a patent, but for our purposes they still provide important additional information as to the sectoral strengths and weaknesses of the advanced countries.

The same data described in Chapter 3 will be analyzed here at the sectoral level, using the database developed by CHI Research for patents granted and patent citations in the US, broken down by 383 three-digit US patent classes and by 43 two-digit US Standard Industrial Classification product fields (SIC). While the former follows the current patent classification of the US Patent Office, the latter is obtained from a concordance between patent classes and SIC classes (US Patent and Trademark Office 1985, 1988). In some cases, all patents of a class are assigned to a specific SIC class; when the same patent class can refer to more than one SIC class, the number of patents is split equally among them. Fractional data occur also when the same patent has inventors from multiple countries, as the number of patents is subdivided according to the number of inventors of each country (See CHI Research 1989a, p. 6).

This Chapter will consider patent counts and patent citations in terms of 43 SIC classes. The same analysis will be carried out as in the previous Chapter, calculating the index of Technological Revealed Comparative Advantage (TRCA) for all sectors and for advanced countries. The only difference from the procedure followed in Chapter 4 is that the shares of each field and the shares of countries have been computed on total patents and patent citations in the US, while in the previous Chapter these shares were obtained from the number of foreign patents only, excluding the domestic patents of the country of registration. As the aim of the analysis of Chapter 4 was to compare patenting patterns in different countries, it was important to avoid the distortion caused by the very high share of domestic patents in national patent offices (US patents in the US, French patents in France, German patents in Germany). Now we consider only one country of registration, the US, and the US pattern of specialization also has to be investigated; therefore the percentages and the indexes of specialization will be obtained by using all patents, including those from US inventors.

5.2. The Sectoral Distribution of Patents and Patent Citations in the US by SIC Classes

An overview of the data used here is provided by Table 5.1,which shows for 43 SIC classes an index of their R&D intensity, the rate of change of the total number of patents between the two periods considered, the percentage distribution of patents granted and patent citations, and the average number of citations per patent. These data, and the analysis of the profile of specialization of each country, focus on two seven-year periods, 1975–81 and 1982–88. The number of years considered in each period ensures that the erratic variations in annual patenting data are avoided, allowing a solid picture of the specialization pattern to emerge for both the late 1970s and the 1980s. Comparison of these two periods also provides a much needed dynamic perspective on the process of technological specialization and further enriches the description of the situation in the 1980s presented in Chapter 4.

In order to identify the different research intensities and the growth performances of the SIC classes considered, two indicators are presented in Table 5.1. The first is the index of R&D intensity of each SIC class, based on the OECD ratios of R&D spending to sales. The classes are grouped into four categories, as defined by a study of the Niedersachisches Institut fur Wirtschaftsforschung in Hannover (Grupp and Legler (1989), for an application see Engelsman and Van Raan (1990)). "Very high tech" classes include Drugs and Medicines, Agricultural and other chemicals, Office computing machines, and Electronic components and communications equipment. "Protected very high tech" classes include those areas – Aircraft, Guided missiles and Ordnance – where government and military procurement are particularly important. Both these classes have R&D expenditure greater than 8% of sales. Classes with an R&D intensity between 3% and 8%

of sales are labelled "Medium high tech", and those with a share below 3% are grouped into the "Medium and low tech" field, combining the two OECD categories with lowest R&D intensities. This typology of SIC classes will allow a better interpretation of the profile of specialization of individual countries resulting from the analysis of their sectoral patenting activities.

The second indicator shown in Table 5.1 is the rate of change of the number of patents between the 1975–81 period and the 1982–88 period. This variable contributes to a better understanding of the importance of each SIC class, differentiating the areas where patenting is growing faster or is declining. The most rapid growth is found in electrical and electronic classes (Office computing machines, Radio TV equipment, Electronic components and communications equipment, etc.), as well as in some mechanical-engineering fields and in Drugs and Medicines. The classes where the number of patents granted is falling include some chemical fields (Organic chemicals, Paints and allied chemicals, etc.) and classes related to raw materials and traditional mechanical activities (Primary ferrous products, Railroad equipment, etc.).

A fast growing patent class can be related to rapidly developing technological advances or to increasing competition among firms in particular sectors. The next Chapter will investigate in detail the growth patterns of a more disaggregated distribution of patents granted in the US, documenting the position of individual countries. This preliminary evidence makes it possible to qualify the specialization profile emerging from sectoral data on patents. The fields of high R&D intensity and fast growth in patenting are generally associated with productions with higher value added and growing markets. A country's strength in these fields is likely to be more relevant than a relative advantage in a class where R&D is low and the number of patents is decreasing.

Comparison of the two indicators so far considered reveals that most "Very high tech" and "Protected very high tech" classes have substantially increased the number of patents across the two periods considered, at rates ranging from 51% in the class of Office computing to 14% in the class of Drugs and Medicines (Ordnance is the exception, with a growth rate below 5%). However, some of the highest rates of increase in patenting are found in classes listed as "Medium high tech", especially in the electric and electronic fields and in Professional instruments. The "Medium low tech classes" generally show stagnant or declining rates of change of the total number of patents in the two periods.

The percentage distribution of patents granted and patent citations across SIC classes highlights the structure of the database and allows comparison between the sectoral patterns shown by patents and patent citations. The last two columns of Table 5.1 show the average number of citations per patent and summarize the information on citation patterns. The differences in the propensity to patent in fields such as chemicals and aircraft have already been pointed out in the previous Chapter, suggesting caution in the comparison of the numbers of patents across technological fields. The same caution should now be exercised in comparing the number of citations received by patents of different SIC classes. A higher propensity to patent in a class can also lead to a higher average number of citations per patent, and citation habits are also likely to differ across sectors due to the structure of the

TABLE 5.1.

THE SECTORAL DISTRIBUTION OF PATENTS AND PATENT CITATIONS IN THE US BY SIC CLASSES
R&D intensities of SIC classes: VH: Very high tech; PH: Protected very high tech; MH: Medium high tech; ML: Medium low tech; Rates of change of patents 1975-81 to 1982-88; percent distribution of patents and patent citations, 1975-81 and 1982-88, and average number of citations per patent

SIC classes	R&D intens. of class	% change of patents	Patents % distribution 75-81	82-88	Citations % distribution 75-81	82-88	Aver.numb.of citat. per patent 75-81	82-88
1 Food,Kindred Products	ML	-9.92%	0.84%	0.70%	0.80%	0.57%	3.25	0.97
2 Textile Mill Products	ML	2.09%	0.70%	0.66%	0.74%	0.75%	3.63	1.38
3 Inorganic Chemicals	MH	-9.35%	1.88%	1.57%	2.01%	1.65%	3.65	1.26
4 Organic Chemicals	ML	-31.03%	7.37%	4.71%	5.33%	3.51%	2.46	0.90
5 Plastic Matrls,Synth Res	MH	3.61%	1.57%	1.51%	1.90%	1.65%	4.11	1.32
6 Agricultural & other chem.	VH	16.73%	1.58%	1.71%	1.11%	1.22%	2.37	0.86
7 Soaps,Detergents,Clnrs	ML	0.10%	0.67%	0.62%	0.88%	0.68%	4.45	1.33
8 Paints,Allied Chemicals	MH	-16.85%	0.78%	0.60%	0.85%	0.61%	3.72	1.23
9 Misc Chemical Products	MH	6.44%	1.03%	1.01%	1.11%	1.15%	3.69	1.37
10 Drugs & Medicines	VH	14.42%	1.96%	2.07%	1.47%	1.56%	2.56	0.91
11 Petrol,Nat Gas Extr,Ref	ML	11.41%	1.15%	1.19%	1.60%	1.55%	4.71	1.58
12 Rubber,Misc Plast Prods	ML	5.89%	4.30%	4.22%	4.92%	4.71%	3.89	1.35
13 Stone,Clay,Glass,Concr	ML	7.42%	1.99%	1.98%	2.11%	2.02%	3.60	1.23
14 Primary Ferrous Prods	ML	-14.28%	0.57%	0.45%	0.49%	0.30%	2.95	0.79
15 Prim,Sec Non-Ferr Prods	ML	0.95%	0.56%	0.53%	0.48%	0.39%	2.90	0.88
16 Fabricated Metal Prods	MH	3.17%	8.23%	7.87%	7.64%	6.58%	3.16	1.01
17 Engines & Turbines	MH	14.59%	1.01%	1.07%	1.06%	1.11%	3.59	1.25
18 Farm,Garden Mach & Equip	ML	-12.52%	1.47%	1.19%	1.17%	0.84%	2.71	0.85
19 Cnstr,Mng,Metal Hand Eqp	MH	-4.39%	2.90%	2.56%	2.31%	2.09%	2.71	0.98
20 Metal Working Mach,Equip	MH	-3.60%	2.71%	2.42%	1.80%	1.54%	2.26	0.77
21 Office Comput,Acctg Mach	VH	51.26%	2.27%	3.18%	3.93%	5.14%	5.88	1.95
22 Spec Ind Mach(exc M Wrk)	MH	-9.26%	5.19%	4.36%	4.24%	3.52%	2.77	0.97
23 Genrl Indust Mach,Equip	ML	3.30%	5.46%	5.22%	4.82%	4.44%	3.00	1.03
24 Refrig,Servc Indust Mach	MH	1.46%	1.54%	1.45%	1.38%	1.19%	3.04	0.99
25 Misc Mach (exc Electric)	MH	5.28%	0.69%	0.67%	0.66%	0.61%	3.23	1.08
26 Electr Trans,Distr Equip	MH	20.93%	2.29%	2.56%	2.40%	2.78%	3.57	1.31
27 Electr Indust Apparatus	MH	23.69%	2.12%	2.43%	2.31%	2.70%	3.70	1.34
28 Household Appliances	MH	13.92%	0.77%	0.81%	0.70%	0.69%	3.12	1.04
29 Electr Lightng,Wirng Eqp	MH	18.14%	0.73%	0.80%	0.72%	0.88%	3.34	1.34
30 Misc Elec Mach,Eqp,Suppl	MH	-6.84%	0.91%	0.78%	1.05%	0.79%	3.93	1.22
31 Radio,TV Receiving Equip	MH	39.94%	0.94%	1.22%	1.23%	1.81%	4.43	1.79
32 Elect Cmp,Acc,Comm Equip	VH	30.56%	9.09%	10.99%	11.78%	14.74%	4.40	1.62
33 Motor Veh,Motor Veh Eqp	MH	17.07%	2.07%	2.24%	2.00%	2.44%	3.29	1.31
34 Guid Mssls,Spce Veh,Prts	PH	19.36%	0.06%	0.06%	0.04%	0.04%	2.49	0.67
35 Ship,Boat Bldng & Repair	ML	-2.52%	0.35%	0.31%	0.23%	0.18%	2.25	0.68
36 Railroad Equipment	ML	-11.89%	0.27%	0.22%	0.18%	0.17%	2.28	0.91
37 Motorcycles,Bicy & Parts	ML	21.65%	0.17%	0.20%	0.16%	0.23%	3.19	1.41
38 Misc Transportation Eqp	ML	-8.80%	0.44%	0.37%	0.33%	0.36%	2.57	1.18
39 Ordnance (exc Missiles)	PH	4.93%	0.53%	0.52%	0.33%	0.34%	2.09	0.80
40 Airc. & other mech. parts	PH	19.72%	1.06%	1.17%	1.10%	1.28%	3.54	1.32
41 Prof,Scien Instruments	MH	21.23%	11.58%	13.00%	13.66%	14.80%	4.01	1.37
42 Unclassified Patents		-	0.03%	1.12%	0.03%	0.11%	3.49	0.12
43 Other Industries		1.21%	8.18%	7.67%	6.94%	6.30%	2.88	0.99
All Prod Flds Combined		7.99%	100.00%	100.00%	100.00%	100.00%	3.40	1.21

Source: CNR-ISRDS elaboration on CHI Research data; R&D intensities from OECD and Grupp-Legler (1989)

technological knowledge in specific fields. With these qualifications, the average number of citations per patent may still highlight the citing behaviour across classes, indicating areas of greater importance in the current pattern of inventive activities. The large number of total patent citations should be noted; more than 1,500,000 citations refer to the 450,000 patents granted in the US in the 1975–81 period; 585,000 citations are found to the 485,000 more recent patents granted in the 1982–88 period. Such a large number of citations makes the statistical analysis of such data reliable and significant.

The number of citations a patent receives from later ones obviously depends on its year of registration; recent patents can be cited by a much smaller number of patents than older ones, and more citations are added for each year that has passed since a patent was registered. The absolute number of citations received and the average number of citations per patent are therefore much lower in the second period considered than in the first. An additional point regarding the use of patent citations is the different "speed" of citing activity across fields, which may lead to an overestimate of the citations per patent in fields where citations spread more rapidly, and to underestimating the impact of patented inventions in classes with slower citing habits. As we are considering a 14-year set of citations, from 1975 to 1988, to patents granted over the same years, grouped into just two seven-years periods, such differences should not affect our data.

These characteristcs of citation data do not affect the profile of sectoral specialization illustrated by the shares of SIC classes and by the indexes of technological advantage measured on patents and citations, which will be examined in the next section. The different sizes of classes and the different propensities to cite across sectors are shown in Table 5.1. Examination of the percentage distribution across SIC classes reveals that the largest share of total patents and patent citations is that of Professional and scientific instruments, which increases in the two periods, 1975–81 and 1982–88, from 11.6% to 13% for patents and from 13.7% to 14.8% for citations. A similarly high and growing share is found in the class of Electric components, rising from 9% to 11% for patents and from 11.8% to 14.7% for citations.

High but decreasing shares are found for Fabricated metal, Organic chemicals, General industrial machinery, Special industrial machinery and Rubber and plastics, with shares ranging from 8% to 4%. A number of electronic-related sectors, with a share of about 2% of all patents and citations, show rapid growth (Office computing, Electrical transmission, Electrical industrial apparatus), and the same pattern is found for Motor vehicles. The classes which increase their share of total patents across the two periods are generally found in the "Very high tech" or "Medium high tech" groups, while the sharpest slumps in the shares of both patents and citations can be found in both "Medium high tech" and "Medium low tech" classes.

The data of Table 5.1 show how different the pattern across SIC classes is, ranging from 5.88 average citations per patent in the class of Office computing, to 2.09 for Ordnance in the first period, and from 1.95 for Office computing to 0.67 for Guided Missiles in the second period. The most highly cited classes in 1975–81 are Office computing, Petroleum and natural gas, Rubber and plastics, Soaps and detergents, Radio and tv receivers and Electrical components. In 1982–88 the list includes Office computing, Radio and TV receivers, Electrical components, Petroleum and natural gas, Motorcycles and bicycles, Professional and scientific instruments and Miscellaneous chemicals. Again, the growing importance of electronics and the decline of chemicals in the pattern of patent citations are evident trends.

5.3. Country Highlights

After this overview across sectors, the next step is the analysis of the specialization indexes already defined in Chapter 4. On the base of these indexes, a ranking of the areas of greater specialization of each country is produced for both patents and citations in the two periods considered. Table 5.2 shows the top five SIC classes with the highest index of technological specialization for major countries. The full rankings of all SIC classes are listed in Table 5.3, showing the specialization profile of each country in patents and patent citations for both periods. Additional caution is called for in interpretation here. Especially for small countries and classes with few patents, even a small number of citations can produce an abnormally high index of specialization and rank position. Careful analysis is therefore required in examining the results.

The profiles obtained for all countries for patent counts and patent citations in both periods are also compared in order to assess the consistency of the specialization profile over time and between patenting data and the impact indicator; the indexes of correlation[1] are shown in Table 5.4. Their comparison over time offers an indication of the degree of technological cumulativeness in patenting sectors, showing how close both the patent and the citation profiles of the late 1970s are to the pattern that emerged in the 1980s. The comparison of patents and patent citations makes it possible to asses how close the picture drawn from patent data is to the more refined profile of national specialization based on the impact of patented inventions. As we shall see, most countries show fairly high correlation coefficients both between patents and citations and over time. Since the number of citations received by each country reflects the number of patents granted, this result is not surprising: as indicators of technological specialization, both patent counts and patent citations provide similar pictures of countries' positions.

The United States. As patents granted in the US are considered here, the specialization profile emerging from the activity of US inventors is marked by what we described in the previous Chapter as the "domestic market effect". About 60% of all patents granted by the US Patent Office are to US residents (62% in 1975–81 and 55% in 1982–88), and their share of patent citations is even higher (65% in 1975–81 and 57% in 1982–88). This suggests that the profile of specialization for the US is closer to the picture of sectoral specialization for the data as a whole and its indexes show much less variation than those of foreign countries. As shown in Chapter 4, the specialization profile of the US in foreign markets is substantially different from that emerging from domestic patenting, and in interpreting the areas of US technological strength a special effort is needed to identify the fields where purely domestic factors are crucial.

Examination of the patent data shows the field of greatest specialization to be

TABLE 5.2.

THE TOP 5 SECTORS OF GREATER TECHNOLOGICAL STRENGTH OF ADVANCED COUNTRIES
INDEXES OF SPECIALIZATION FOR PATENTS GRANTED AND PATENT CITATIONS, 1975-81 AND 1982-88

Country	Indicator	Top 5 SIC classes with the highest specialization indexes for patents and patent citations 1	2	3	4	5
USA	Pat.75-81	Petrol,Nat Gas Extr.	Guid Miss.Space Veh.	Ordnance	Farm,Garden Machin.	Fabric. Metal Prod.
	Pat.82-88	Petrol,Nat Gas Extr.	Guid Miss.Space Veh.	Farm,Garden Machin.	Electr Lighting	Food,Kindr. Products
	Cit.75-81	Petrol,Nat Gas Extr.	Guid Miss.Space Veh.	Farm,Garden Machin.	Ordnance	Misc Chemical Prod.
	Cit.82-88	Petrol,Nat Gas Extr.	Guid Miss.Space Veh.	Farm,Garden Machin.	Ship,Boat Building	Food,Kindr. Products
JAPAN	Pat.75-81	Radio,TV Rec. Equip.	Motorcycles & Parts	Engines & Turbines	Primary Ferrous Prod.	Aircr,Oth.Mech.Parts
	Pat.82-88	Radio,TV Rec. Equip.	Office Comput Mach.	Motorcycles & Parts	Motor Vehicles	Engines & Turbines
	Cit.75-81	Radio,TV Rec. Equip.	Misc Machin.	Motorcycles & Parts	Aircr,Oth.Mech.Parts	Engines & Turbines
	Cit.82-88	Motorcycles & Parts	Motor Vehicles	Aircr,Oth.Mech.Parts	Engines & Turbines	Radio,TV Rec. Equip.
W.GERMANY	Pat.75-81	Organic Chemicals	Misc Elec Machin.	Special Ind Machin.	Agric. & Other Chem.	Plastic Materials
	Pat.82-88	Special Ind Machin.	Ordnance	Railroad Equipment	Organic Chemicals	Metal Working Machin.
	Cit.75-81	Special Ind Machin.	Motor Vehicles	Aircr,Oth.Mech.Parts	Organic Chemicals	Unclassified Patents
	Cit.82-88	Railroad Equipment	Special Ind Machin.	Aircr,Oth.Mech.Parts	Ordnance	Motor Vehicles
UN.KING.	Pat.75-81	Agric. & Other Chem.	Soaps,Detergents	Motorcycles & Parts	Drugs & Medicines	Guid Miss.Space Veh.
	Pat.82-88	Soaps,Detergents	Guid Miss.Space Veh.	Agric. & Other Chem.	Drugs & Medicines	Aircr,Oth.Mech.Parts
	Cit.75-81	Agric. & Other Chem.	Soaps,Detergents	Drugs & Medicines	Misc Elec Machin.	Organic Chemicals
	Cit.82-88	Soaps,Detergents	Agric. & Other Chem.	Drugs & Medicines	Railroad Equipment	Non-Ferrous Products
FRANCE	Pat.75-81	Drugs & Medicines	Agric. & Other Chem.	Guid Miss.Space Veh.	Aircr,Oth.Mech.Parts	Ship,Boat Building
	Pat.82-88	Guid Miss.Space Veh.	Railroad Equipment	Soaps,Detergents	Aircr,Oth.Mech.Parts	Drugs & Medicines
	Cit.75-81	Guid Miss.Space Veh.	Motorcycles & Parts	Aircr,Oth.Mech.Parts	Drugs & Medicines	Ship,Boat Building
	Cit.82-88	Guid Miss.Space Veh.	Railroad Equipment	Soaps,Detergents	Ship,Boat Building	Aircr,Oth.Mech.Parts
CANADA	Pat.75-81	Ship,Boat Building	Non-Ferrous Products	Misc Transportation	Petrol,Nat Gas Extr.	Farm,Garden Machin.
	Pat.82-88	Ship,Boat Building	Farm,Garden Machin.	Misc Transportation	Non-Ferrous Products	Food,Kindr. Products
	Cit.75-81	Ship,Boat Building	Non-Ferrous Products	Misc Transportation	Petrol,Nat Gas Extr.	Special Ind Machin.
	Cit.82-88	Farm,Garden Machin.	Ship,Boat Building	Food,Kindr. Products	Electr Lighting	Misc Transportation
ITALY	Pat.75-81	Household Appliances	Special Ind Machin.	Drugs & Medicines	Agric. & Other Chem.	Organic Chemicals
	Pat.82-88	Household Appliances	Agric. & Other Chem.	Special Ind Machin.	Drugs & Medicines	Railroad Equipment
	Cit.75-81	Household Appliances	Plastic Materials	Special Ind Machin.	Motorcycles	Misc Transportation
	Cit.82-88	Special Ind Machin.	Household Appliances	Agric. & Other Chem.	Drugs & Medicines	Constr,Mining Equip.

TABLE 5.2 (continued).

THE TOP 5 SECTORS OF GREATER TECHNOLOGICAL STRENGTH OF ADVANCED COUNTRIES
INDEXES OF SPECIALIZATION FOR PATENTS GRANTED AND PATENT CITATIONS, 1975-81 AND 1982-88

Country	Indicator	Top 5 SIC classes with the highest specialization indexes for patents and patent citations 1	2	3	4	5
NETHERL.	Pat.75-81	Radio,TV Rec. Equip.	Electr Lighting	Elect Comp,Comm. Eq.	Food,Kindr. Products	Farm,Garden Machin.
	Pat.82-88	Radio,TV Rec. Equip.	Food,Kindr. Products	Elect Comp,Comm. Eq.	Soaps,Detergents	Farm,Garden Machin.
	Cit.75-81	Radio,TV Rec. Equip.	Elect Comp,Comm. Eq.	Farm,Garden Machin.	Food,Kindr. Products	Ship,Boat Building
	Cit.82-88	Guid Miss.Space Veh.	Food,Kindr. Products	Radio,TV Rec. Equip.	Soaps,Detergents	Elect Comp,Comm. Eq.
SWITZERL.	Pat.75-81	Organic Chemicals	Agric. & Other Chem.	Textile Products	Drugs & Medicines	Food,Kindr. Products
	Pat.82-88	Agric. & Other Chem.	Organic Chemicals	Textile Products	Special Ind Machin.	Ordnance
	Cit.75-81	Organic Chemicals	Agric. & Other Chem.	Drugs & Medicines	Textile Mill Products	Food,Kindr. Products
	Cit.82-88	Agric. & Other Chem.	Organic Chemicals	Special Ind Machin.	Textile Mill Products	Drugs & Medicines
SWEDEN	Pat.75-81	Primary Ferrous Prod	Ordnance	Constr,Mining Equip.	Ship,Boat Building	Non-Ferrous Products
	Pat.82-88	Ship,Boat Building	Ordnance	Railroad Equipment	Primary Ferrous Prods	Guid Miss.Space Veh.
	Cit.75-81	Ordnance	Primary Ferrous Prod	Constr,Mining Equip.	Metal Working Machin.	Non-Ferrous Products
	Cit.82-88	Ordnance	Guid Miss.Space Veh.	Primary Ferrous Prod	Non-Ferrous Products	Ship,Boat Building
BELGIUM	Pat.75-81	Soaps,Detergents	Plastic Materials	Primary Ferrous Prods	Non-Ferrous Products	Drugs & Medicines
	Pat.82-88	Soaps,Detergents	Non-Ferrous Products	Farm,Garden Machin.	Inorganic Chemicals	Textile Mill Products
	Cit.75-81	Soaps,Detergents	Plastic Materials	Non-Ferrous Products	Textile Mill Products	Drugs & Medicines
	Cit.82-88	Soaps,Detergents	Non-Ferrous Products	Drugs & Medicines	Electr Lighting	Inorganic Chemicals
SPAIN	Pat.75-81	Ship,Boat Building	Ordnance	Special Ind Machin.	Food,Kindr. Products	Drugs & Medicines
	Pat.82-88	Ordnance	Special Ind Machin.	Misc Transportation	Engines & Turbines	Ship,Boat Building
	Cit.75-81	Ordnance	Ship,Boat Building	Special Ind Machin.	Misc Transportation	Drugs & Medicines
	Cit.82-88	Special Ind Machin.	Fabric. Metal Prod.	Ordnance	Electr Lighting	Railroad Equipment
DENMARK	Pat.75-81	Food,Kindr. Products	Gen. Industr Machin.	Household Appliances	Stone,Clay,Glass	Misc Machin.
	Pat.82-88	Food,Kindr. Products	Farm,Garden Machin.	Drugs & Medicines	Gen. Industr Machin.	Engines & Turbines
	Cit.75-81	Food,Kindr. Products	Drugs & Medicines	Agric. & Other Chem.	Gen. Industr Machin.	Prof,Scient. Instr.
	Cit.82-88	Food,Kindr. Products	Drugs & Medicines	Agric. & Other Chem.	Textile Mill Products	Gen. Industr Machin.

Source: CNR-ISRDS elaboration on CHI Research data

TABLE 5.3.

RANKS OF THE INDEXES OF TECHNOLOGICAL SPECIALIZATION
FOR PATENTS GRANTED AND PATENT CITATIONS IN THE USA BY SIC CLASSES
Period I: 1975-81
Period II: 1982-88

SIC classes	UNITED STATES				JAPAN				W. GERMANY				UNITED KINGDOM				FRANCE			
	Patent		Citat.		Patent		Citat.		Patent		Citat.		Patent		Citat.		Patent		Citat.	
	I	II	I	II	I	II	I	II	I	II	I	II	I	II	I	II	I	II	I	II
1 Food,Kindred Products	11	6	10	6	26	34	30	37	42	42	42	40	19	24	10	20	38	23	38	27
2 Textile Mill Products	32	31	28	29	15	14	15	16	6	16	13	17	9	11	11	10	25	19	29	19
3 Inorganic Chemicals	26	14	25	11	19	31	21	33	13	17	23	24	16	28	15	27	10	7	20	18
4 Organic Chemicals	36	26	34	24	23	26	22	28	1	4	4	9	13	15	5	8	14	12	13	20
5 Plastic Matrls,Synth	31	20	32	22	9	13	8	15	5	14	7	14	34	42	27	23	28	41	31	40
6 Agric. & other Chem.	41	36	39	33	28	28	26	32	4	8	9	13	1	3	1	2	2	6	6	10
7 Soaps,Detergents,Clnr	17	12	16	21	31	38	36	38	27	22	30	29	2	1	2	1	9	3	15	3
8 Paints,Allied Chemica	20	15	20	15	14	19	12	20	16	12	16	18	33	30	33	30	41	42	40	43
9 Misc Chemical Product	8	8	6	14	21	21	25	24	36	35	38	33	32	12	32	16	36	28	34	24
10 Drugs & Medicines	42	30	38	25	16	25	18	29	11	23	19	19	4	4	3	3	1	5	4	7
11 Petrol,Nat Gas Extr,R	1	1	1	1	40	40	41	40	43	43	43	42	41	43	40	43	40	39	42	42
12 Rubber,Misc Plast Pro	18	16	18	19	17	17	16	14	23	21	24	22	25	27	26	29	31	38	32	39
13 Stone,Clay,Glass,Conc	22	22	24	18	18	16	17	18	29	26	27	30	8	10	8	15	19	14	17	14
14 Primary Ferrous Prods	43	38	43	35	4	9	6	12	17	27	10	27	20	17	12	13	11	17	19	9
15 Prim,Sec Non-Ferr Pro	39	35	31	28	11	11	14	13	32	32	32	35	18	6	31	5	7	10	7	37
16 Fabricated Metal Prod	7	7	7	12	37	32	34	30	34	30	33	25	35	33	35	33	33	25	30	25
17 Engines & Turbines	35	41	36	40	3	5	5	4	14	13	6	6	11	7	16	21	18	29	12	26
18 Farm,Garden Mach & Eq	6	4	4	4	41	41	40	41	40	34	37	26	37	36	38	41	32	35	27	21
19 Cnstr,Mng,Metal Hand	15	9	17	13	39	39	37	36	26	20	25	20	14	9	17	12	24	33	18	32
20 Metal Working Mach,Eq	28	27	26	26	29	27	29	22	10	5	15	10	27	35	24	34	27	31	28	28
21 Office Comput,Acctg M	19	34	15	30	6	2	11	7	33	40	39	39	38	41	36	42	30	43	26	33
22 Spec Ind Mach(exc M W	38	37	37	36	27	29	28	27	3	1	1	2	24	22	30	17	15	15	9	13
23 Genrl Indust Mach,Equ	25	28	27	31	25	22	24	17	15	10	18	12	26	19	28	19	29	18	22	16
24 Refrig,Servc Indust M	13	10	11	10	33	33	33	35	24	25	26	28	31	25	37	28	35	32	35	31
25 Misc Mach (exc Electr	30	32	41	39	7	12	2	6	19	7	8	7	23	18	34	11	17	26	14	29
26 Electr Trans,Distr Eq	12	17	12	17	20	15	19	21	30	24	34	21	30	16	23	6	22	11	24	11
27 Electr Indust Apparat	33	33	30	34	10	7	9	9	20	28	21	23	22	37	21	38	13	22	25	23
28 Household Appliances	21	29	19	27	22	18	23	19	18	11	20	11	40	39	41	39	42	36	41	22
29 Electr Lightng,Wirng	9	5	8	9	36	30	35	34	31	36	31	31	17	34	22	24	26	16	23	8
30 Misc Elec Mach,Eqp,Su	27	21	21	16	24	20	31	23	2	15	12	15	7	23	4	25	23	27	33	30
31 Radio,TV Receiving Eq	40	39	33	38	1	1	1	5	37	38	35	36	28	38	20	26	43	37	39	35
32 Elect Cmp,Acc,Comm Eq	14	23	13	20	12	8	13	11	38	37	41	38	29	29	29	37	12	13	21	15
33 Motor Veh,Motor Veh E	29	42	35	42	13	4	7	2	9	6	2	5	15	26	19	32	16	24	8	17
34 Guid Mssls,Spce Veh,P	2	2	3	2	43	42	43	43	25	19	22	41	5	2	7	7	3	1	1	1
35 Ship,Boat Bldng & Rep	16	13	14	5	35	37	38	39	35	39	36	34	21	14	14	35	5	9	5	4
36 Railroad Equipment	24	25	29	32	30	36	27	26	7	3	11	1	10	8	13	4	6	2	10	2
37 Motorcycles,Bicy & Pa	37	43	42	43	2	3	3	1	22	18	14	8	3	21	9	14	8	20	2	34
38 Misc Transportation E	10	18	22	37	32	23	20	8	39	31	28	16	12	13	18	18	21	30	11	41
39 Ordnance (exc Missile	4	11	5	8	42	43	42	42	21	2	17	4	42	32	42	22	20	8	16	6
40 Aircr.&Oth.Mech.Parts	34	40	40	41	5	6	4	3	12	9	3	3	6	5	6	9	4	4	3	5
41 Prof,Scien Instrument	23	24	23	23	8	10	10	10	28	33	29	32	36	20	25	31	37	34	36	36
42 Unclassified Patents	5	19	2	3	38	24	39	25	8	29	5	43	43	31	43	40	34	21	43	12
43 Other Industries	3	3	9	7	34	35	32	31	41	41	40	37	39	40	39	36	39	40	37	38

Source: Elaboration on CHI Research data

TABLE 5.3 (continued).

RANKS OF THE INDEXES OF TECHNOLOGICAL SPECIALIZATION
FOR PATENTS GRANTED AND PATENT CITATIONS IN THE USA BY SIC CLASSES
Period I: 1975-81
Period II: 1982-88

SIC classes	ITALY				NETHERLANDS				BELGIUM				DENMARK				SPAIN			
	Patent		Citat.		Patent		Citat.		Patent		Citat.		Patent		Citat.		Patent		Citat.	
	I	II	I	II	I	II	I	II	I	II	I	II	I	II	I	II	I	II	I	II
1 Food,Kindred Product	35	25	32	21	4	2	4	2	28	16	27	33	1	1	1	1	4	15	10	9
2 Textile Mill Product	16	31	26	33	27	24	24	11	10	5	4	9	18	19	29	4	24	38	31	32
3 Inorganic Chemicals	9	19	10	6	13	14	15	28	7	4	8	5	31	8	20	7	23	20	30	16
4 Organic Chemicals	5	6	9	9	11	11	10	10	12	19	14	13	16	20	8	12	12	26	19	26
5 Plastic Matrls,Synth	6	12	2	10	17	9	17	6	2	21	2	27	41	41	41	41	41	41	41	37
6 Agric. & other Chem.	4	2	8	3	23	34	29	26	8	7	12	8	11	6	3	3	7	7	11	15
7 Soaps,Detergents,Cln	37	39	30	39	7	4	6	4	1	1	1	1	20	28	27	26	36	39	37	25
8 Paints,Allied Chemic	29	27	20	25	29	31	30	30	14	14	13	17	37	32	37	32	38	32	32	41
9 Misc Chemical Produc	40	41	38	42	20	32	31	33	37	8	30	6	38	35	38	27	37	35	39	30
10 Drugs & Medicines	3	4	6	4	28	30	16	15	5	6	5	3	6	3	2	2	5	6	5	17
11 Petrol,Nat Gas Extr,	42	42	42	41	16	20	11	23	32	27	37	24	40	34	40	37	40	43	40	39
12 Rubber,Misc Plast Pr	14	21	12	26	21	18	28	16	9	13	9	11	22	17	17	14	29	24	21	19
13 Stone,Clay,Glass,Con	28	26	35	28	15	13	20	8	16	9	11	10	4	13	10	22	26	27	28	34
14 Primary Ferrous Prod	11	37	21	34	31	28	32	39	3	10	6	16	26	31	25	25	18	29	7	27
15 Prim,Sec Non-Ferr Pr	26	36	36	32	35	38	35	41	4	2	3	2	29	37	18	28	28	23	13	31
16 Fabricated Metal Pro	31	29	28	22	22	23	26	24	22	29	25	26	19	12	16	11	19	14	17	2
17 Engines & Turbines	27	30	19	24	34	39	37	40	40	39	41	39	21	5	21	16	14	4	18	14
18 Farm,Garden Mach & E	34	28	37	14	5	5	3	9	11	3	7	7	13	2	24	8	13	13	20	21
19 Cnstr,Mng,Metal Hand	17	10	15	5	8	15	8	21	23	26	35	28	12	10	11	17	20	30	14	20
20 Metal Working Mach,E	12	9	16	8	32	26	34	31	18	23	24	14	24	22	14	24	25	25	24	29
21 Office Comput,Acctg	7	22	7	11	14	22	13	20	29	34	33	25	36	36	36	35	35	36	33	36
22 Spec Ind Mach(exc M	2	3	3	1	12	12	14	7	15	11	17	15	8	9	9	18	3	2	3	1
23 Genrl Indust Mach,Eq	10	11	11	7	19	27	19	27	19	18	20	12	2	4	4	5	16	11	15	18
24 Refrig,Servc Indust	13	14	18	19	9	16	9	14	24	20	29	20	14	7	22	10	27	17	23	22
25 Misc Mach (exc Elect	18	18	17	16	37	33	38	38	36	31	38	32	5	16	23	20	22	16	29	23
26 Electr Trans,Distr E	30	32	22	31	18	17	23	12	31	36	28	34	10	29	13	30	31	34	27	35
27 Electr Indust Appara	24	33	25	35	24	21	22	29	30	32	26	30	7	18	12	6	32	18	35	33
28 Household Appliances	1	1	1	2	26	29	27	34	25	28	18	29	3	11	7	21	21	21	9	8
29 Electr Lightng,Wirng	20	15	29	12	2	7	7	18	26	17	15	4	30	26	31	31	17	19	25	5
30 Misc Elec Mach,Eqp,S	41	40	41	40	10	8	18	17	17	38	23	35	28	14	19	9	39	22	38	24
31 Radio,TV Receiving E	22	38	24	37	1	1	1	3	35	25	32	19	25	39	28	36	33	40	36	40
32 Elect Cmp,Acc,Comm E	32	35	31	30	3	3	2	5	33	30	36	23	33	33	35	33	34	37	34	38
33 Motor Veh,Motor Veh	19	17	14	17	38	35	39	36	27	41	31	36	32	30	34	34	10	9	12	10
34 Guid Mssls,Spce Veh,	43	43	40	43	43	10	43	1	43	43	43	43	27	42	30	42	42	42	42	42
35 Ship,Boat Bldng & Re	39	24	39	29	6	6	5	13	39	33	39	41	23	25	26	15	1	5	2	11
36 Railroad Equipment	38	5	27	20	33	43	25	43	20	35	22	38	39	43	39	40	6	28	22	6
37 Motorcycles,Bicy & P	8	7	4	36	40	41	41	22	38	42	19	40	34	40	32	43	9	33	6	12
38 Misc Transportation	15	16	5	23	41	37	40	32	34	37	34	31	17	27	6	38	11	3	4	7
39 Ordnance (exc Missil	23	8	34	18	42	42	33	42	6	12	10	21	43	23	42	13	2	1	1	4
40 Aircr.&Oth.Mech.Part	21	23	13	13	39	40	42	37	41	40	40	37	35	38	33	39	15	10	16	13
41 Prof,Scien Instrumen	33	34	33	27	25	25	21	25	13	24	16	22	9	15	5	19	30	31	26	28
42 Unclassified Patents	36	13	43	38	30	19	12	19	42	15	42	42	42	24	43	29	43	8	43	43
43 Other Industries	25	20	23	15	36	36	36	35	21	22	21	18	15	21	15	23	8	12	8	3

TABLE 5.3 (continued).

RANKS OF THE INDEXES OF TECHNOLOGICAL SPECIALIZATION
FOR PATENTS GRANTED AND PATENT CITATIONS IN THE USA BY SIC CLASSES
Period I: 1975-81
Period II: 1982-88

SIC classes	IRELAND				PORTUGAL				GREECE				SWITZERLAND				SWEDEN				CANADA			
	Patent		Citat.		Patent		Citat.		Patent		Citat.		Patent		Citat.		Patent		Citat.		Patent		Citat.	
	I	II	I	II	I	II	I	II	I	II	I	II	I	II	I	II	I	II	I	II	I	II	I	II
1 Food,Kindred Product	4	25	17	11	40	36	28	29	36	37	30	31	5	7	5	13	22	18	23	20	19	6	20	4
2 Textile Mill Product	43	17	34	35	36	28	29	37	3	42	42	42	3	3	4	4	36	29	35	26	38	36	39	31
3 Inorganic Chemicals	9	24	20	26	32	33	23	39	13	3	22	2	41	35	35	37	37	28	38	17	13	15	13	14
4 Organic Chemicals	34	29	30	22	10	2	8	5	24	6	19	9	1	2	1	2	40	40	36	39	37	41	30	39
5 Plastic Matrls,Synth	7	1	6	1	30	22	22	15	25	5	12	12	15	26	15	24	42	43	42	43	42	42	41	42
6 Agric. & other Chem.	27	23	40	5	20	9	16	2	17	30	14	40	2	1	2	1	32	36	26	29	33	38	36	32
7 Soaps,Detergents,Cln	40	22	32	16	31	6	34	1	38	7	31	3	8	8	16	10	34	38	31	35	43	43	43	43
8 Paints,Allied Chemic	32	33	39	34	35	26	27	41	11	1	28	10	11	15	13	9	38	41	40	42	36	37	27	38
9 Misc Chemical Produc	35	19	33	21	33	14	40	4	43	43	43	43	29	32	30	30	31	35	33	30	27	22	31	17
10 Drugs & Medicines	33	10	31	2	6	7	20	3	14	12	4	16	4	6	3	5	28	33	10	27	29	34	34	23
11 Petrol,Nat Gas Extr,	26	32	21	30	14	23	6	25	32	9	29	1	42	42	42	42	41	39	41	40	4	8	4	12
12 Rubber,Misc Plast Pr	23	6	24	3	21	5	18	9	10	32	16	18	24	27	25	26	30	30	34	33	23	24	24	26
13 Stone,Clay,Glass,Con	16	13	22	18	29	8	32	20	5	31	9	36	33	30	33	31	16	23	17	23	25	26	22	21
14 Primary Ferrous Prod	38	41	38	32	43	43	43	43	1	10	1	23	23	21	17	27	1	4	2	3	7	14	11	10
15 Prim,Sec Non-Ferr Pr	42	42	42	42	38	32	33	22	4	14	2	37	19	14	19	14	5	6	5	4	2	5	2	9
16 Fabricated Metal Pro	21	7	27	8	22	11	21	12	7	29	10	24	25	23	24	17	9	12	9	12	12	11	15	11
17 Engines & Turbines	22	26	18	20	7	10	7	6	12	2	36	28	27	31	29	36	8	17	13	13	28	30	32	34
18 Farm,Garden Mach & E	3	2	7	27	28	15	37	30	8	23	5	38	9	22	8	25	15	15	18	14	6	2	7	1
19 Cnstr,Mng,Metal Hand	19	8	15	17	8	29	9	33	20	26	11	4	21	18	18	12	3	8	3	7	8	7	8	13
20 Metal Working Mach,E	24	20	25	7	39	17	41	38	18	36	15	29	13	11	12	8	6	7	4	6	17	18	17	16
21 Office Comput,Acctg	25	9	43	14	11	25	1	23	23	39	23	20	17	28	26	29	23	34	27	36	41	39	40	36
22 Spec Ind Mach(exc M	12	18	23	13	12	19	4	24	6	28	3	39	6	4	6	3	12	13	7	9	9	10	6	7
23 Genrl Indust Mach,Eq	28	27	29	24	16	16	17	7	16	21	7	17	14	13	11	7	7	11	6	8	14	21	18	19
24 Refrig,Servc Indust	10	11	8	19	41	12	30	13	31	33	39	34	18	19	21	22	11	10	15	16	10	12	10	8
25 Misc Mach (exc Elect	14	38	12	37	9	31	3	26	19	35	18	41	30	24	32	33	10	20	21	28	21	25	28	35
26 Electr Trans,Distr E	15	12	16	33	27	41	38	32	30	25	21	33	22	16	22	23	29	32	32	34	22	23	19	20
27 Electr Indust Appara	31	4	11	12	18	18	35	11	29	19	20	27	10	9	10	11	14	19	22	19	32	32	25	25
28 Household Appliances	13	31	26	23	24	13	24	8	22	18	40	11	20	12	14	16	19	9	11	11	15	17	16	15
29 Electr Lightng,Wirng	5	3	3	15	37	30	31	18	41	38	32	21	28	29	27	20	26	31	30	32	11	9	12	5
30 Misc Elec Mach,Eqp,S	36	28	35	41	3	40	36	28	26	40	13	22	38	38	38	28	13	14	14	10	30	35	29	28
31 Radio,TV Receiving E	39	36	37	25	17	34	2	36	39	34	25	35	34	37	36	35	39	42	39	41	39	40	38	40
32 Elect Cmp,Acc,Comm E	29	16	28	10	23	20	15	17	15	27	6	15	31	34	28	34	35	37	37	31	24	27	23	18
33 Motor Veh,Motor Veh	8	30	13	31	13	38	12	19	28	13	27	19	32	36	37	38	20	21	24	18	16	19	21	24
34 Guid Mssls,Spce Veh,	2	43	2	43	25	35	26	27	35	41	38	32	43	43	43	43	33	5	19	2	18	33	9	30
35 Ship,Boat Bldng & Re	37	40	36	28	26	1	39	34	2	17	34	6	37	40	31	21	4	1	8	5	1	1	1	2
36 Railroad Equipment	30	37	9	29	1	39	11	40	37	16	26	5	12	10	9	15	17	3	16	38	20	16	14	37
37 Motorcycles,Bicy & P	6	34	4	36	4	37	10	21	40	11	41	8	35	41	41	41	21	27	25	37	34	20	33	41
38 Misc Transportation	17	5	5	4	2	42	14	31	33	20	33	13	39	33	40	32	25	22	20	21	3	3	3	6
39 Ordnance (exc Missil	41	35	41	38	34	24	25	35	34	4	24	25	7	5	7	6	2	2	1	1	40	31	35	27
40 Aircr.&Oth.Mech.Part	18	39	14	40	5	27	5	16	21	8	37	30	36	39	39	40	24	26	29	24	35	29	37	33
41 Prof,Scien Instrumen	20	15	19	6	19	21	19	14	27	22	17	14	16	20	20	18	27	25	28	25	31	28	26	22
42 Unclassified Patents	1	21	1	39	42	3	42	42	42	15	35	26	40	17	34	39	43	24	43	22	26	13	42	29
43 Other Industries	11	14	10	9	15	4	13	10	9	24	8	7	26	25	23	19	18	16	12	15	5	4	5	3

TABLE 5.4.

CORRELATION COEFFICIENTS ACROSS TECHNOLOGICAL SPECIALIZATION PROFILES FOR PATENTS GRANTED AND PATENT CITATIONS IN THE US, 1975-81 AND 1982-88

Correlation coefficients between indexes of technological specialization of advanced countries by SIC classes

Countries	Correlation coefficients between: Patents 1975-81 and Citations 1975-81	Patents 1982-88 and Citations 1982-88	Patents 1975-81 and Patents 1982-88	Citations 1975-81 and Citations 1982-88
USA	0.95	0.93	0.88	0.86
Japan	0.95	0.93	0.93	0.91
West Germany	0.92	0.90	0.86	0.84
United Kingdom	0.88	0.88	0.78	0.56
France	0.91	0.87	0.79	0.77
Canada	0.92	0.88	0.85	0.70
Italy	0.95	0.91	0.89	0.81
Netherlands	0.96	0.78	0.87	0.47
Switzerland	0.98	0.95	0.95	0.93
Sweden	0.94	0.82	0.84	0.81
Belgium	0.91	0.94	0.74	0.80
Spain	0.86	0.69	0.74	0.59
Denmark	0.87	0.91	0.77	0.75
Ireland	0.91	0.74	0.10 *	0.08 *
Portugal	0.30 *	0.28 *	-0.14 *	-0.12 *
Greece	0.66	0.41	0.05 *	-0.19 *

Source: Elaborations on CHI Research data, supplied to ISRDS-CNR

Correlation coefficients across 41 SIC classes. The residual classes 'Other Industries' and 'Unclassified' have been excluded.

All coefficients are statistically significant at the 5% level, except the coefficients marked with an asterisk (*)

Petroleum and natural gas, followed at a distance by Guided missiles, Ordnance, Farm and garden machinery and Fabricated metals. Due to the nature of domestic patenting, there is also a large number of Unclassified patents and those assigned to Other industries. Food and kindred products and Electric lighting show an increase in the index from the late 1970s to the 1980s, while Inorganic chemicals and Plastic materials go from a relative weakness to a small relative advantage over the two periods.

On the other hand, Office computing moves from a small relative specialization in the first period to a despecialization in the second. A worsening of US technological weakness can be found in Motor vehicles, Aircraft and other mechanical parts, Engines and turbines, Motorcycles and bicycles. In the 1980s, Motor vehicles and Motocycles are the two sectors of greatest US technological weakness. The picture emerging from patent citations is extremely close to that of patents, both in the list of strengths and weaknesses and in the pattern of change over the two periods.

These areas of US strength in the domestic market tend to be concentrated in military-related sectors of "very high tech" where government procurement is crucial; in fields related to the exploitation of natural resources; in some mechanical and electrical classes. The two latter fields are part of the "medium and low tech" group and in all these classes the total number of patents has either been falling or increasing slowly. These fields are often similar to the sectors of relative US advantage that emerged from the analysis of patents classified by IPC classes in the previous Chapter. Paint and petroleum, Weapons, Agriculture, Foodstuff, Electricity were IPC classes showing a relative US strength confirmed by the evidence from the SIC classification.

But how different is the US specialization profile in its domestic market from that resulting from patenting abroad? A lesson may be drawn here from the comparison in the previous Chapter (Table 4.4) between the specialization indexes based on patents in the US and at the EPO. In the latter, major US strengths emerged in Computing, Organic compounds and pharmaceuticals, fields where SIC data show a general US weakness; these are likely to be cases of underestimating a country's strength when only domestic patenting activity is considered.

Finally, Table 5.4 shows the correlation coefficients between the specialization profiles of patents and citations for both periods. The data for the US show a very high correlation between patents and citations (coefficients of 0.95 and 0.93 in the two periods). Over time, however, the specialization profiles show some change, with a lower correlation coefficient (0.88 and 0.86) for patents and citations across the two periods.

Japan. The specialization profiles for patent counts and patent citations in the two periods are very similar also for Japan, which always shows high correlation coefficients both between the two variables and over time, in spite of the near doubling of its share of patents granted in the US (from 10% in 1975–81 to 18% in 1982–88) and of patent citations (from 12% to 21%). The variation in the specialization indexes is higher, with greater distance between the areas of strength and weakness. The field of greatest specialization is by far Radio and TV receivers, but its importance decreases rapidly in the two periods and is smaller if we look at patent citations. In the 1980s, the highest specialization measured on citations is found in Motorcycles.

Areas of fast growing strength for both patents and citations are Office computing, Aircraft and other mechanical parts[2], Engines and turbines. Together with Motorcycles, Motor vehicles and Non-electrical machines, the latter two fields, show values for citation indexes which are much higher that patent counts, suggesting an important impact of Japanese inventions in these clusters of mechanical and electronic fields. Japan's greatest weakness for both patents and citations is found in many petrochemical sectors and in military-related areas such as Ordnance and Guided missiles, where there is a very low absolute number of patents.

The fields of Japanese strength are concentrated in electrical, electronic and mechanical sectors characterized by scale intensive production of consumer durable and intermediate goods, or by the specialized production of machinery. The fields based on natural resources, science intensive and military-related sectors are those where Japan shows a relative weakness. The areas of Japanese advantage include a spectrum of different R&D intensities, from "very high tech" to "medium and low tech" classes, but almost all of them are sectors of rapid growth in the total number of patents granted in the US, suggesting an intense innovative activity and a concentration of efforts in the fields of emerging technological importance.

West Germany. The areas of greatest German strength are the chemical sectors (Organic chemicals, Agricultural and other chemicals and Plastic materials), showing high values for both patents and citations, with some decrease over the two periods. Other areas have registered a fast growth and are now at the top of the specialization list; these include Specialized industrial machinery, Ordnance, Railroad equipment, Metal working machinery, Motor vehicles, Motorcycles and Engines and turbines. They show much higher values for citations than for patents, indicating that these are areas of greater impact of German inventions.

Scale intensive production in chemical and mechanical sectors are the major strengths of German technological specialization, with relative advantages also in the specialized production of machinery and in a few science intensive and "protected very high tech" activities. German weaknesses can be found in information intensive technologies (Office computing and other electronic-related sectors) and in classes related to natural resources (Petroleum and natural gas). The fields of German strength are mostly in "medium high tech" and "medium low tech" classes, which show a variety of patterns of growth and decline of total patenting activity.

The correlation coefficients display a close but falling relation between the specialization profiles shown by patents and citations (coefficients of 0.92 in the first period and 0.90 in the second). Across the two periods, citations show greater changes in the pattern of specialization, with a coefficient of 0.84.

France. The areas of fast growing French strength include Guided missiles and space vehicles, Aircraft and other mechanical parts, and Ordnance in a cluster of military-related and state-supported industries, marked however by a generally small absolute number of patents. Marked specialization is also found in Drugs and medicines and increasing specialization in Railroad equipment (another state-supported sector) and Soaps and detergents. Positive but falling specialization occurs in Motor vehicles, Engines and turbines and in various chemical fields. In electronic-related sectors, France shows a stable strength in Electronic components and communications equipment. French technological strengths include fields of "protected very high tech" where military procurement is crucial, as well as science intensive sectors (such as pharmaceuticals), and classes where the scale of production (in chemical and mechanical fields) is an important factor. Many areas of French strength lie in classes with a fast growing number of total patents, either in the "very high tech" or in the "medium-high tech" group. The correlation coefficients indicate a close similarity between the specialization profiles resulting from patents and citations (0.88 in the both periods), and much less continuity in the country's strengths over time, especially for citations (the coefficient across the two periods is 0.56).

United Kingdom. The major area of British strength appears to be Soaps and detergents, where there is an increase over the two periods considered and an extremely high value (3.94) for the citation index of the second period. Table 5.1 has already shown that this sector has a particularly high average number of citations per patent for the world's total. Citation data increasing over time and much higher than patent data can be found also for Agricultural and other chemicals, Drugs and medicines, Non ferrous products, Electrical transmission. Guided missiles (although with a very small number of patents) shows a high and growing index. Marked drops in the specialization index can be found in Electrical machinery, Motor vehicles, Aircraft and other mechanical parts, and in some chemical fields; the greatest weakness is (somewhat surprisingly) in Petroleum and natural gas, with particularly low citation values, and in electrical and electronic classes. The areas of strength include some science intensive (chemicals and pharmaceuticals) and military related fields falling into the "very high tech" and "protected very high tech" classes, as well as sectors with scale intensive production (chemical and mechanical fields). While the former include classes with a growing number of total patents, the latter are generally classes in "medium and low tech" sectors with a falling number of total patents.

Data for British technological specialization show some variations over time (the correlation coefficients across the two periods is 0.79 and 0.77 for both patents and citations), while a close similarity is found between patents and citations in both periods.

Italy. Household appliances, Agricultural and other chemicals[3], Special industrial machinery and Drugs and medicines are by far the areas of greatest Italian strength for both patents and citations. Other sectors of positive but decreasing specialization include Organic chemicals and Plastic materials. Some growing strength can be found in Construction, General industrial machinery and Motor vehicles. High indexes for citations are found in some machinery and chemical classes, as well as in Office computing, where a marked fall is seen in Italy's activity over the two periods. Areas of Italian weakness include most electronic-related sectors as well as Petroleum and natural gas.

The classes of Italian specialization tend to lie in the specialized production of machinery (in a variety of fields related to mechanical engineering) and in the scale intensive production of consumer durables (in the electrical and mechanical fields) and chemical products. Contrary to the results of the previous Chapter, based on the IPC classification of patents, no clear strength emerges in traditional industries such as textile and food. Such results should be related to the different criteria followed by the two classifications. While the IPC includes the industrial machinery with the final products of each sector, the SIC separates them, identifying Italy's strengths in the areas of Specialized industrial machinery. The fields of Italian strength show widely different degrees of research intensities and rates of change of total patenting in the US. The correlation coefficients suggest a close similarity between the sectoral distribution of patents and citations and a greater variability over time of the specialization profile based on citations.

Canada. Canada's greatest technological specialization is in Shipbuilding, Miscellaneous transportation and in areas such as Non-ferrous and Ferrous products, and Farm and garden machinery. The latter is the only field where the index for citations increases over time, and areas of strengths emerge from citation data in Food and kindred products, Electrical lighting, Special industrial machinery. Areas of weakness include electronic-related sectors and all chemical fields. The classes of Canadian strength are often in resource-related fields, with some specialized engineering production; they are generally "medium and low tech" classes with falling numbers of total patents in the US. The correlations are always high, both between patents and citations and over time, across the two periods considered.

The Netherlands. The pattern of specialization of the Netherlands shows clear and stable strengths in Radio and TV receivers, Electrical components, Food and kindred products, Soaps and detergents. Important fields showing a major decline are Electrical lighting, Farm and garden machinery and Shipbuilding. Rapid growth is shown in Plastic materials, with a particularly high value for citations (an abnormally high value of citations in 1982–88 in Guided missiles is caused by an extremely low absolute number of citations). The relative Dutch specialization in electric and electronic areas is a rare aspect among European countries, and it is evident for both scale intensive consumer durables and for the specialized production of intermediate goods. Other strengths are in the traditional agricultural sector and in scale intensive chemical technologies. The country's advantages are in classes showing varying degrees of R&D intensities and growth rates of patenting. The correlation coefficients between patents and citations drop from 0.96 in the first period to 0.78 in the second, and the stability over time of the specialization profile for citations is also very low (coefficient of 0.47).

Belgium. Chemicals are by far Belgium's area of greatest technological specialization, with very high indexes for all the relevant SIC classes, but with diverging patterns over time and between patents and citations. Soaps and detergents present the highest value for both patents and citations in both periods. Citation data are very high, almost twice the indexes for patents and both variables double their specialization measure from the late 1970s to the 1980s. Higher values for citations than for patents and a growing specialization can also be found in Drugs and medicines, Miscellaneous chemicals and Inorganic chemicals. A sharp worsening of Belgian specialization in Plastic materials can be found over the two periods. Other areas of growing strength include Electric lighting, Farm and garden machinery and Non ferrous products. Fields of weakness can be found in the mechanical and electronic technologies. The classes of Belgian strength tend to be in scale intensive technologies belonging to the "medium and low tech" group, with falling numbers of total patents in the US. The sectoral distributions of patents and citations show a strong and even growing correlation over the two periods (indexes of 0.91 and 0.94), and a high stability over time of the specialization profiles can also be found.

Switzerland. The specialization profile of Switzerland is more stable and consistent, with the greatest and growing strengths in Agricultural and other chemicals, Organic chemicals, Textile products and Special industrial machinery. Other relative advantages are found in Food and kindred products, Soaps and detergents and Drugs and medicines, which show declining strength. Areas of weakness again include most mechanical, electronic and aerospace sectors. The areas of Swiss specialization include science intensive areas of "very high tech" (chemical and pharmaceutical), scale intensive chemical technologies, specialized production of machinery and traditional industries such as textiles, lying in "medium and low tech" fields. The correlation coefficients between patents and citations are very high (0.98 and 0.95 in the two periods) and little change can be found over time in the country's specialization profiles.

Sweden. Ferrous, Non ferrous and Fabricated metal products, Metal working machinery, and General industrial machinery are the major cluster of Swedish specialization. Other related strengths can be found in Construction and mining, Shipbuilding and Ordnance. A weakness in chemicals and some electronic sectors complete the country's specialization profile. The classes of Swedish strength are generally in the specialized production of machinery and resource related sectors, lying in "medium and low tech" areas, with falling numbers of total patents. There are high correlation coefficients between patents and citations and a fairly stable picture over time in the specialization profiles.

Denmark. The agroindustrial system is clearly the greatest strength of Denmark, which shows very high and growing indexes of specialization in Food and kindred products, Agricultural and other chemicals, Farm and garden machinery, Refrigerating and service industries. Citation indexes are generally higher than for patent data, and for Food products they double from the first to the second period. Drugs and medicines, General industrial machinery and Household appliances are other areas of strength, while the fields of greatest weakness include some other chemical and mechanical sectors electronics and aerospace. The cluster of Danish specialization is, with few exceptions, in the "medium and low tech" sector, in classes showing a falling number of total patents in the US. Comparison of patents and citations shows that the correlation coefficients are high and present a stable pattern of specialization over time.

Spain. A cluster of mechanical sectors emerges as the major area of Spanish technological specialization, including Special and General industrial machinery, Motor vehicles and Fabricated metal. A few other areas of apparent strength have a very low absolute number of patents and citations, and cannot be considered fields of relative advantage. Chemical and electronic classes show the major weaknesses of the country, whose strengths tend to be in "medium high tech" areas. The specialization profiles for patents and citations diverge over time and show some variation across the two periods especially for citation data.

Ireland. The very low number of patents granted and patent citations for Ireland leads to an erratic profile of specialization, with a very dispersed distribution.

Plastic materials, Farm and garden machinery, Electric lighting emerge as the areas of more substantial strength, while in many classes of chemical, mechanical and electronic sectors there is no significant patenting activity.

Portugal. The only sector where a high index of specialization is obtained from a substantial number of patents and citations for Portugal is Organic chemicals. In the other classes the absolute numbers are too low to produce a reliable indicator of technological strength and weakness.

Greece. Fabricated metal and some other mechanical classes are the only areas of significant specialization for Greece, which has too few patents and citations in most classes to allow a significant analysis of its technological profile.

Overview. At the end of this survey of the strengths and weaknesses of the advanced countries, it is possible to identify some general trends of the specialization indexes of patents and citations. Firstly, the specialization profiles emerging from patents and patent citations are very similar for most countries. While increasing differences can be found in the second period considered, this may be due to the smaller number of citations received by the patents granted in the more recent period. Secondly, while patent data show a strong stability over time in pointing out the areas of national advantage, the sectoral distribution of patent citations tends to change more rapidly. While the importance of technological cumulativeness is evident, the relative positions of classes do change, especially in the areas with the greatest impact of inventive activities. This suggests that the impact indicator provides a more selective picture of national strengths and weaknesses and is more responsive to the changes occurring in the outcome of technological activities of advanced countries.

The results of this analysis, which is based on patenting in the US only, integrate the findings of the previous Chapter on patenting at the major international institutions. Data based on patents in the US have made it possible to examine changes over time in the sectoral patterns of specialization and to identify the more specific country profile resulting from an indicator of the impact that patents have. Combination of these new insights into the sectoral structure of the advanced countries' inventive activities with the extensive comparative analysis of the previous Chapter makes it possible to paint a full and detailed picture of the strengths and weaknesses of individual countries as measured by the patent indicator. The next Chapter will offer a more detailed analysis which may qualify the different nature and impact of the specialization patterns here reviewed. It will examine the subfields of fastest growth in patented inventions and describe the countries' position in the technologies where the greatest innovative activity and competition are found.

Notes

1. The correlation coefficients are calculated on the indexes for 41 SIC classes, excluding the two residual classes of Other industries and Unclassified patents. The same procedure was followed

in the previous Chapter, excluding the IPC class of Unclassified patents.

2. The SIC class "Aircraft and parts" includes an important group of mechanical parts, and we have renamed the class accordingly. The work by Patel and Pavitt (1987a), which uses a classification making it possible to identify the two parts of this class, shows that Japanese strength in this class is due to activity in "Other mechanical parts", and not in "Aircraft".
3. The SIC class "Agricultural chemicals" includes a variety of other chemical products and has been renamed accordingly. Italy's strength in this class is due to activity in the "Other chemicals" group rather than in specific agricultural products, as shown by Patel and Pavitt (1987a).

CHAPTER 6

National Technological Specialization and Sectoral Growth Rates in Patenting

6.1. Patent Classes and the Technological Race

The previous two Chapters have mapped the different countries' sectoral strengths and weaknesses in technological activity by using patent data as an indicator. We have already stressed that the quality of individual patents is highly skewed, and that while some have a dramatic economic impact, many patents never become actual innovations. In the previous Chapter we also considered an indicator of their impact, i.e. patent citations, and significant cross-country differences were shown to exist in the average impact of patented inventions at the aggregate level. At the sectoral level, however, a strong positive correlation emerged between the specialization profiles measured on the number of patents and of the citations received.

The description given in Chapters 4 and 5 of the different countries' specializations (based on patenting) already contains an assessment of their technological capabilities. We have pointed out the importance of sectoral advantages in high (or pervasive) technologies, and the problems of a specialization limited to traditional fields. A more specific evaluation of the relevance of individual sectors in the growth of technological and economic activities, however, has still to be carried out.

Since all fields do not have the same relevance in technological and economic activities, this Chapter addresses the question of the "quality" of national patterns of technological specialization. We will examine how the particular mix of a country's sectoral strengths and weaknesses contributes to national technological performance and how it compares with the international sectoral patterns of innovations. For this purpose, a methodology able to identify the role played by each sector in the innovative processes has been developed.

The rates of change shown by the total number of patents in each class are used as an indicator of their relative technological importance. A fast growing patent class often indicates original developments in scientific and technological knowledge, but can also be related to increasing competition in, or diffusion of already known inventions and innovations. Several case studies (see, among others, Walsh (1984),

Wheal and McNally (1986), Achilladelis et al. (1987), Trajtenberg (1990)) have shown that fast growing patent classes are generally associated with increasing competition among firms or countries for leadership in selected technological areas and are likely to correspond to expanding markets. In more general terms, it could be argued that the fields of rapid patent expansion are at the technological frontier, and will represent the common technologies of future economic systems.

The classifications used in the previous Chapters are not detailed enough to identify the areas of greatest technological dynamism. Growth rates have already been calculated for the SIC classes used in the previous Chapter, but we are aware that in a classification at the 2 digit level, sub-fields of rapid expansion are often merged with sub-fields with stagnant or declining rates of change. A pioneering study on fast growing patent classes (Patel and Soete, 1988), for example, focused on the 300 fastest growing sub-classes out of a total of 110,000. For our purposes, such a detailed level of disaggregation is not needed. In fact, national positions will be here assessed in all the classes and not only in the fast growing sub-set, thus requiring a less disaggregated classification.

6.2. The Rates of Change in Patent Classes at the 3 Digit Level

This Chapter will consider the 118 technological fields defined by the 3 digit International Patent Sub-Classes focusing on patents granted in the US alone for two different reasons. On the one hand, it is not possible to compute sectoral growth rates on the basis of the European Patent Office data since the rapid growth of applications due to the relatively recent foundation of this institution has affected each class in a different way. On the other, the results obtained in Chapter 4 show that the US patent system provides reliable information on technological specialization for the majority of countries. The most notable exception is represented by the US itself, since we are considering its patents in the domestic market. The findings for this country should therefore be considered with some caution.

Table 6.1 reports the growth rates in each of the 118 IPC sub-classes[1] calculated between the total of patents granted in the 1975–78 period and the 1985–88 period. We have focused on the two extreme four-year periods in order to identify the most important technological transformations. Growth rates were computed on the total number of patents registered in the US by all countries, since we assume that technological developments occur on a global scale. Between the two periods considered, a growth occurred for total patents equal to 11.9%. The contribution of individual classes to this trend is very different, as shown in Table 6.1. Variations in the rates of change across classes at the 3 digit level (according to the IPC classes) are much higher than those at the 2 digit level (which have already been shown in Chapter 5 according to the SIC classification).

The 118 classes have been divided into four quartiles, each of them including classes with the following rates of change:

1st quartile (declining): below –10% (30 classes);
2nd quartile (stagnant): from –10% to 6% (29 classes);
3rd quartile (medium growing): from 6% to 26% (29 classes);
4th quartile (fast growing): above 26% (30 classes).

TABLE 6.1.

RATES OF CHANGE OF PATENTS IN THE US - IPC 3 DIGIT CLASSIFICATION
Rates of change between the 1975-78 and 1985-88 periods

4th QUARTILE FAST GROWING		3rd QUARTILE MEDIUM GROWING		2nd QUARTILE STAGNANT		1st QUARTILE DECLINING	
Crystal growth	(1)	Basic electronic cir	25.4%	Crush/pulv/disintegr	5.6%	Machines/engines for	-10.1%
Disposal of solid wa	485.3%	Sewing;embroidering;	24.3%	Machines or engines	5.3%	Checking devices	-10.7%
Instrument details	169.2%	Weapons	23.7%	Signalling	5.3%	Furnaces;kilns;ovens	-12.4%
Bookbinding;albums;f	101.5%	Basic electric eleme	23.0%	Cleaning	3.4%	Phys/chem process/ap	-13.1%
Information storage	94.0%	Musical instr;acoust	22.3%	Brushware	2.7%	Ammunition;blasting	-13.3%
Biochem;beer;mut/gen	89.6%	Printng;lining mach;	21.8%	Sports;games;amuseme	1.4%	Making paper article	-13.6%
Computing;calculat;c	88.4%	Workng/preserv wood;	19.6%	Life-saving;fire-fig	1.0%	Foods or foodstuffs	-14.1%
Optics	77.6%	Gen/conv/distr of el	19.0%	Hand & travel articl	0.8%	Metallurgy;ferrous/n	-14.4%
Lighting	66.5%	Butcherng;meat,poul,	18.3%	Heatng;ranges;ventil	0.3%	Fluid-pressure actua	-14.6%
Electric communicati	62.7%	Earth drilling;minin	17.9%	Separ solids from so	0.0%	Skins;hides;pelts;le	-14.9%
Med & vet scien;hygi	58.5%	Casting;powder metal	16.4%	Ships/waterborne ves	-0.3%	Dyes;paints;polishes	-16.8%
Refrig/cool;manuf of	53.1%	Drying	15.2%	Building	-0.5%	Metallurgy of iron	-18.0%
Layered products	51.7%	Gen/transmit mech vi	14.9%	Grinding;polishing	-0.6%	Fertilisers	-18.3%
Footwear	49.9%	Furnit;domestic art/	14.8%	Liquid handling	-0.8%	Centrifugal app/mach	-18.4%
Hand tools;port pow	46.7%	Combustion apparatus	14.4%	Conveying;packing;st	-1.5%	Working of plastics	-20.1%
Ropes;cables,not ele	45.8%	Engineer elements/un	14.3%	Electolytic/electoph	-2.4%	Haberdashery;jewelry	-20.7%
Steam generation	44.9%	Baking;edible doughs	14.1%	Petroleum,gas & coke	-3.0%	Workng cement/clay/s	-21.7%
Electr tech not prov	44.6%	Combustion engines	13.2%	Sep solid mat using	-3.3%	Decorative arts	-23.3%
Nuclear physics;nucl	42.0%	Writing,drawing appl	12.6%	Const of road/rail/b	-4.5%	Hand cuttng tools;cu	-24.0%
Pos-displac mach for	38.6%	Aircraft;aviation;co	11.9%	Hoisting;lifting;hau	-5.3%	Presses	-24.0%
Educating;crypto;dis	36.5%	Cements;ceramics;ins	11.2%	Workng/treatmt of me	-5.3%	Yarns;warping or bea	-26.4%
Controlling;regulati	34.1%	Tobacco;cigars;cigar	9.9%	Weaving	-5.8%	Hydraulic eng;founda	-26.7%
Wearing apparel	32.7%	Photog;cinematog;ele	9.7%	Water supply;sewerag	-7.2%	Braiding;lace-makng;	-29.0%
Machine tools	31.3%	Land vehicles for tr	9.0%	Org macromolec compo	-7.4%	Inorganic chemistry	-29.8%
Measuring;testing	30.1%	Treatment of water,w	8.7%	Paper-makng;prod of	-7.7%	Railways	-30.7%
Doors/windows/shutte	29.9%	Vehicles in general	8.7%	Nat/artif threads/fi	-7.8%	Sugar or starch indu	-30.9%
Heat exchange in gen	29.4%	Glass,mineral,slag w	7.0%	Agric;forest;anim hu	-9.3%	Organic chemistry	-33.2%
Spraying/atomising	29.1%	Locks;keys;wind/door	6.9%	Animal & veget oils,	-9.8%	Treat of textiles;la	-40.4%
Saddlery;uphostery	27.4%	Storng/dist gases/li	6.2%	Mechanical metal-wor	-9.9%	Horology	-42.1%
Headwear	26.6%					Explosives;matches	-47.1%
All IPC's Combined	11.14%						

Source: CNR-ISRDS elaboration on CHI Research data

(1): This new class was introduced after the first period.

The ranking of the patent classes according to their rates of growth raises the question of the accuracy of the identification of the sectors of greater technological change. A check on an independent source has been possible, using the detailed classification of high technology products developed by ENEA on the basis of the expert opinion of engineers and technicians (see Amendola and Perrucci, 1988)[2]. The majority of the high technology products defined according to this method are in fact to be found in fast growing patent classes.

The fastest growing fields of patenting include many classes related to microelectronics, such as Information storage, Computing, Optics, Electric communications, Controlling, Measuring & testing. In the declining classes, on the contrary, Organic chemicals and Inorganic chemicals are to be found. In more general terms,

TABLE 6.2.

SHARE OF PATENT CLASSES IN TOTAL US PATENTS CLASSIFIED ACCORDING TO THEIR RATES OF CHANGE, 1975-78 AND 1985-88

	1975-78	1985-88
1st QUARTILE DECLINING CLASSES	21.57%	14.57%
2nd QUARTILE STAGNANT CLASSES	23.92%	20.82%
3rd QUARTILE MEDIUM GROWING CLASSES	29.77%	31.10%
4th QUARTILE FAST GROWING CLASSES	24.60%	33.34%
ALL IPC'S COMBINED	100.00%	100.00%

Source: CNR-ISRDS elaboration on CHI Research data

Note: The four groups are based on the 3 digit IPC subfields divided in quartiles according to their rates of change. The classes of each group are listed in table 6.1.

areas related to the cluster of electronic technologies have grown at higher rates than mechanical and chemical clusters. Fast growing classes also contain some classes generally considered to be traditional products, such as Footwear and Headwear. It should borne in mind, however, that the IPC classes include both products and processes in the same category; the results emerging will appear less surprising when it is recalled that a substantial proportion of the patents in Footwear, for example, relates to specialized industrial machinery, and that Headwear includes helmets. Table 6.2 shows the share of the total number of patents included in each quartile for the periods 1975–78 and 1985–88. The fast growing classes accounted for less than a quarter of all patents in the period 1975–78, but account for a third in 1985–88.

6.3. Country Results

We assume that for each country it is an advantage to hold a higher share of patents in the sectors which are growing above average[3]. In order to assess the position of individual countries, two different measures were developed:

a) The first is the share of a country's patents lying in the fast growing classes in both periods considered (columns 1 and 3 of Table 6.3). By definition, the share has increased for all countries and the results are to be compared to the world average share in the two periods. We have therefore calculated national specialization indexes for the whole of the group including the fast growing classes (columns 2 and 4 of Table 6.3). The index is the TRCA index described in Chapter 4. A substantial specialization in fast growing classes can be found only for Japan in the second period, while in both periods the share of patents held by the US,

TABLE 6.3.

COUNTRIES' TECHNOLOGICAL SPECIALIZATION AND RATES OF CHANGE OF PATENT CLASSES
Patents granted in the US, 3 digit IPC classes

Countries	1 Patents in fast 1975-78 % of tot. patents	2 growing sectors 1975-78 Index of special.	3 1985-88 % of tot. patents	4 1985-88 Index of special.	5 Correlation coefficients between TRCA indexes 1985-88 and rates of growth of patent classes between 1975-78 and 1985-88 (1)
United States	25.52%	1.04	33.69%	1.01	0.09
Japan	24.19%	0.98	37.10%	1.11	0.22
EEC 12	22.87%	0.93	28.50%	0.85	-0.22
FR Germany	20.01%	0.81	27.19%	0.82	-0.27
France	26.63%	1.08	34.68%	1.04	-0.04
United Kingd.	25.87%	1.05	34.73%	1.04	-0.07
Italy	21.63%	0.88	28.37%	0.85	0.00
Netherlands	25.51%	1.04	33.44%	1.00	0.14
Belgium	18.30%	0.74	27.68%	0.83	-0.13
Denmark	27.90%	1.13	32.94%	0.99	-0.14
Spain	19.15%	0.78	23.82%	0.71	-0.17
Ireland	20.25%	0.82	40.60%	1.22	0.15
Portugal	5.56%	0.23	0.00%	0.00	-0.15
Greece	19.84%	0.81	25.95%	0.78	-0.05
Canada	21.88%	0.89	28.78%	0.86	-0.15
Switzerland	18.06%	0.73	26.75%	0.80	-0.25
Sweden	26.18%	1.06	32.29%	0.97	-0.17
Australia	21.88%	0.89	25.99%	0.78	-0.15
World	24.60%	1.00	33.34%	1.00	

Source: CNR-ISRDS elaboration on CHI Research data

(1) The correlation coefficients have been calculated on 116 classes, excluding the class Crystal growth, introduced after the first period considered, and the class Disposal of solid waste, which has very few patents and an extremely high growth rate.

the UK, France and the Netherlands in fast growing classes is only sligthtly above the world average, their specialization index being just above 1. However, this measurement does not take into account the countries' performance in the patent classes belonging to the other quartiles.

b) The second measure compares the rates of growth of patent classes (already shown in Table 6.1) with the index of specialization (the TRCA index) of a country's patenting activity calculated for the same 118 subclasses. The correlation coefficient between the two vectors[4] indicates how close the specialization pattern of a country is to the sectoral distribution of fast growing classes in world patenting. A positive correlation suggests that the country is positively specialized in areas of faster growth and negatively specialized in fields of stagnant of falling patenting. The reverse is true for a negative value of the correlation coefficient.

The results presented in column 5 of Table 6.3 show that only Japan has a sectoral pattern of specialization which exhibits a clear positive relation to the rates of growth of world patenting. The US has a positive correlation, but very close to zero. Among European countries, Germany and Switzerland have the strongest

Countries' position in fast growing, medium growing, stagnant and declining technologies, 3 digit IPC patent classes, rates of change 1975-78 to 1985-88

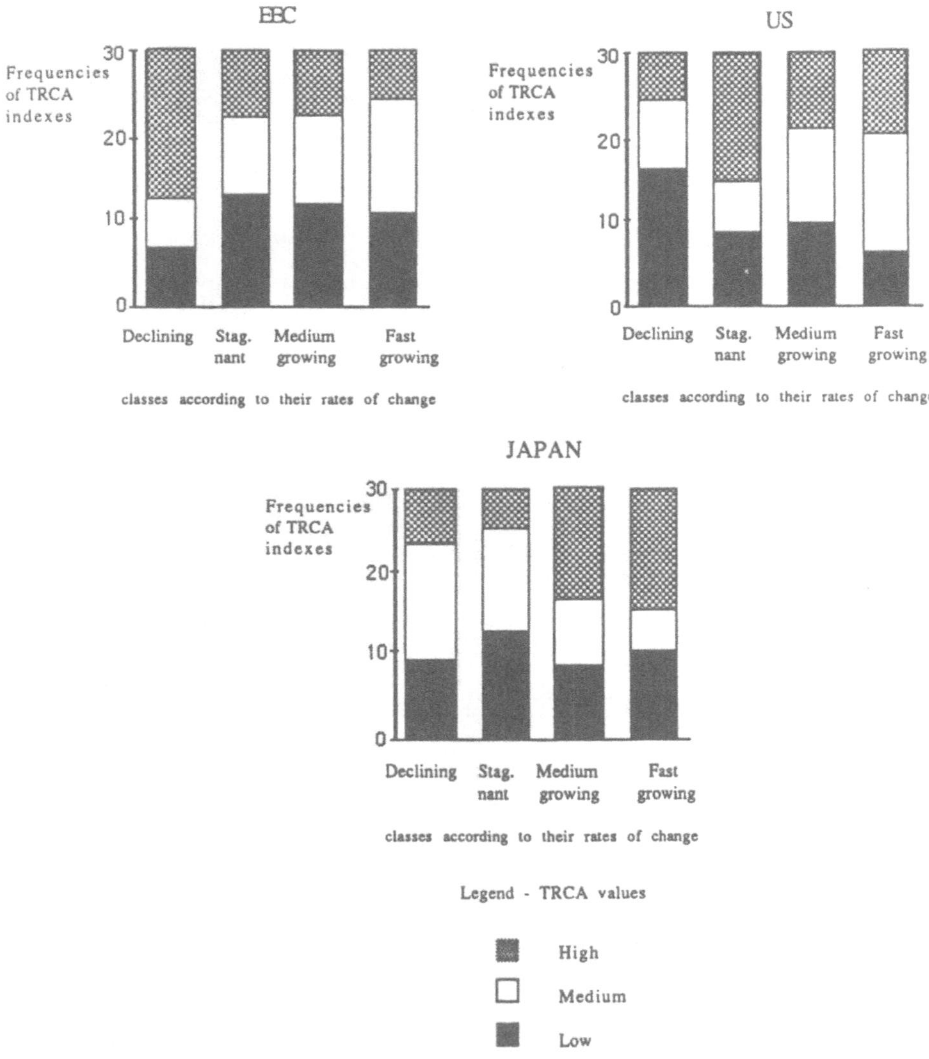

Fig. 6.1. (first part)

negative correlation between their sectoral specialization and the pattern of growth of world patenting; in other words they concentrate their relative strengths in subclasses where world patenting is stagnant or declining. For the Netherlands, on the contrary, there is a positive correlation coefficient, largely due to the activities of this country in fields related to consumer electronics.

This evidence can be documented in greater detail with the graphic represen-

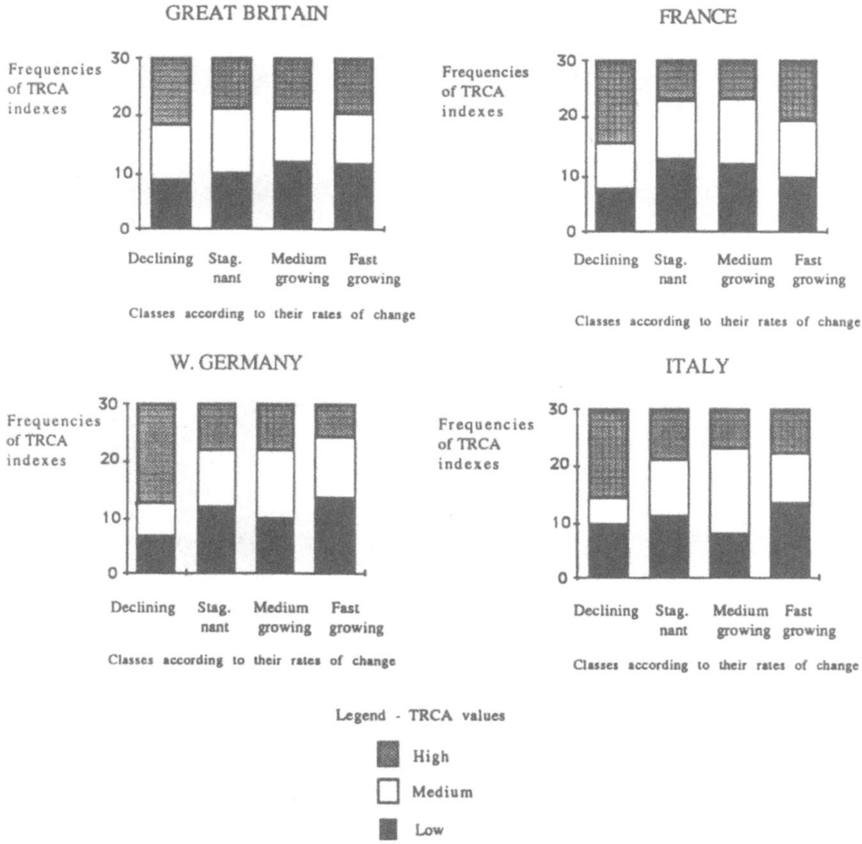

Fig. 6.1. (second part)

tation of Fig. 6.1. For each country the TRCA indexes for the 118 subclasses were calculated for the more recent period, 1985–88; the indexes were then ranked and divided into three groups of 39 classes each: high, medium and low TRCAs. This information is combined in Figure 6.1 with the distribution in four quartiles of the 118 subclasses according to the growth rates in world patenting. Each bar corresponds to a quartile, and is subdivided according to the frequencies of high, medium or low TRCA indexes found for those subclasses. The figure shows where a country distributes its relative strengths (top layer) and weakness (bottom layer), in fields of patenting ranging from fast growing (right bar) to declining (left bar).

The EEC has a below average share in fast growing classes in both the first and the second period. Its position has however weakened over time. In the most recent period, high specialization is found in only 6 (out of 30) of the fast growing classes. Conversely, the EEC tends to be more specialized in the declining classes (18 out of 30 classes). The opposite pattern is shown by Japan. In the first period considered, Japan's share of patents in the fast growing classes was slightly lower than its share of total patents. However, it then reversed its unfavorauble specialization and now

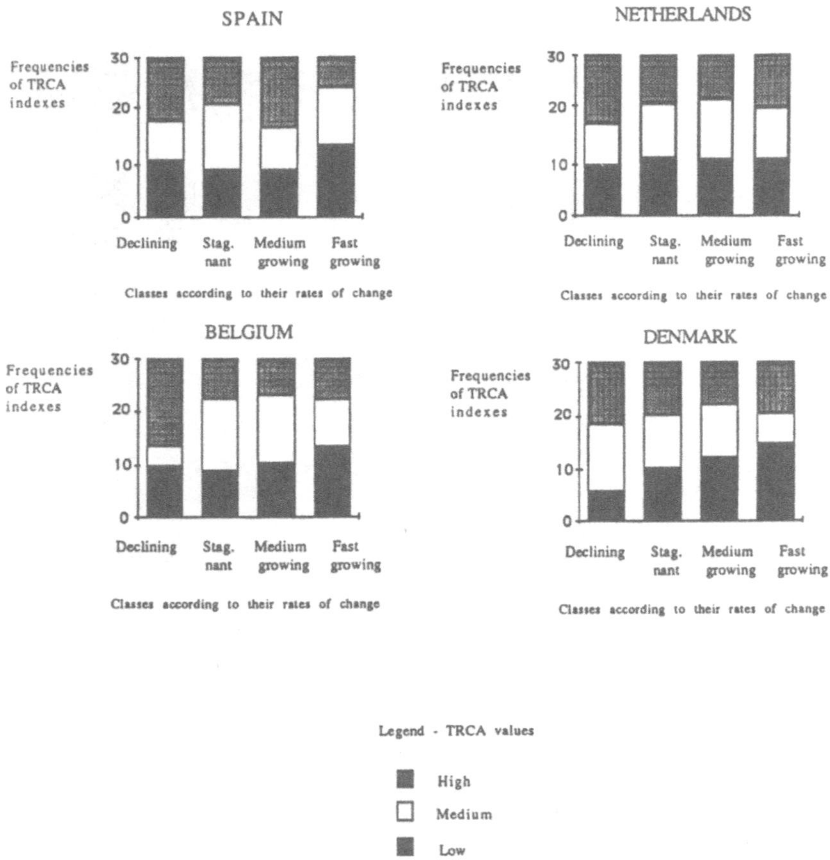

Fig. 6.1. (third part)

shows a high specialization in half of the fast growing classes, while in declining classes it has high TRCAs indexes in only 7 classes out of 30.

The results for the US are biased by the fact that we are here considering its patents in the domestic market. In both periods, however, the US share in fast growing classes is slightly above its total share of patents. In 10 (out of 30) fast growing classes the US presents high TRCA values, and is highly specialized in only 6 of the declining classes. This result is somewhat counter-intuitive: as we have seen in Chapter 3, the total number of patents granted to US residents dropped markedly over the period 1975–88, and one could expect that substantial technological activities of the US (and therefore its sectoral strengths) might be found in classes with a declining rate of change. This result shows that US patents have declined especially in the less dynamic sectors, while it has preserved its technological activities in the fields of more intense competition.

The results shown by the EEC are the outcome of differentiated tendencies across individual countries. Since Germany is the largest European patenting country, the aggregate result reflects German patenting activity more than that of any other

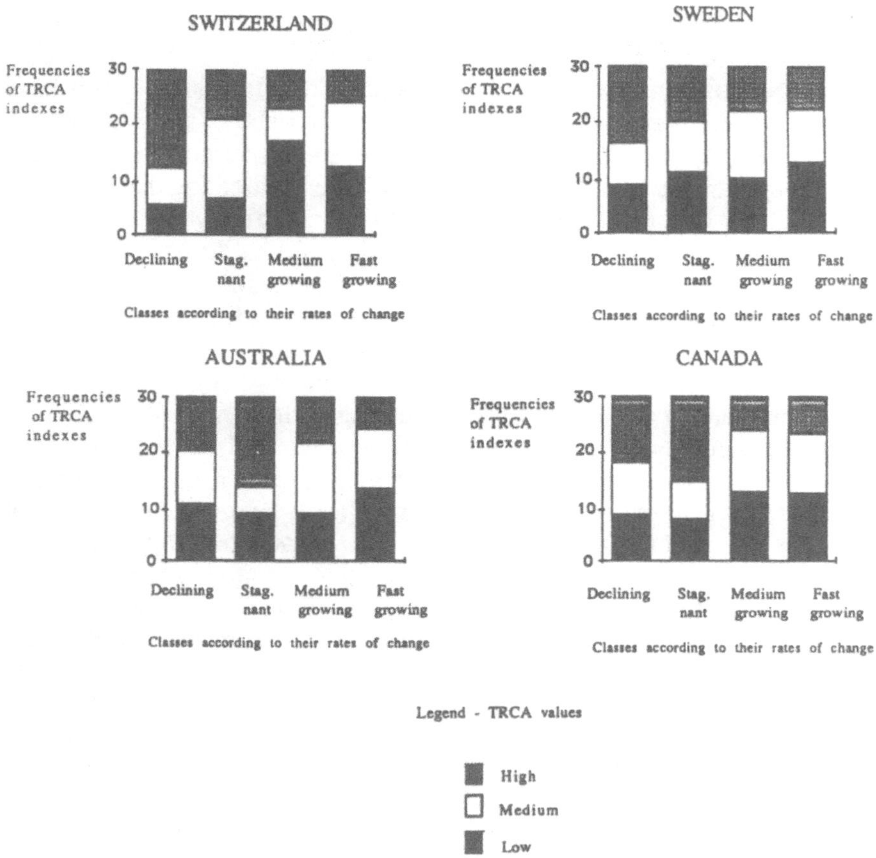

Fig. 6.1. (fourth part)

country. In fact, the sectors of German strengths in the fast growing classes are the same as those of the EEC as a whole. In both periods, the German share of patents in the fast growing classes is below its total share of patents. The United Kingdom, on the opposite, shows a robust specialization in fast growing classes, in spite of the reduction in the total number of patents granted in the US over the last two decades. As for the US, it could be argued that the reduction in the absolute number of patents has affected declining classes more than the fast growing. The third largest European country, France, has a share in fast growing classes slightly above its share of total patents, although it lost ground over the two periods.

6.4. Overview

The comparison of national positions in patent classes ranked according to their rates of change provides some additional information to the mapping of countries' strengths and weaknesses presented in Chapters 4 and 5. This method might help to identify the likely future national performances in technological activities. Sectoral strengths in fast growing classes provide an advantage for future innovations since the relative share of fast growing patent classes increases over time. Consequently, countries specialized in fast growing classes are likely to increase their share of patents at the aggregate level. The findings presented above allow a comparison to be made between the EEC and the Japanese specialization profiles. Japan is consistently specialized in fast growing patent classes, indicating that the sustained growth in patents experienced over the last 20 years is likely to continue also in the future. Conversely, European technological strengths do not lie in the more dynamic sectors and it is likely that the share of European patents will somehow decrease in the future. It was also shown that the leading European technological power, Germany, is not specialized in the most dynamic sectors. While the British total share of technological activities shows a considerable drop, the UK appears to have concentrated its efforts in the fast growing sectors.

Finally, assessment of the US specialization profile is biased by the fact that patenting in its domestic market was considered. However, the fact that several sectoral strengths are achieved in fast growing classes suggests that the declining total amount of patenting is somehow linked to a concentration of innovations in the most dynamic sectors. This would imply that a restructuring of innovative activities has occurred in the US and may suggest a recovery of the total share of US patented inventions. The results emerging in this Chapter should be compared with the aggregate trends in the resources devoted to S&T shown in Chapter 3, where it was shown that Japanese S&T activities have increased much faster than in the EEC countries. This Chapter provides another warning about European performance which can be identified behind the aggregate picture: the current sectoral strengths and weaknesses of Europe and Japan are likely to lead to an increasing gap between the technological performances of the two regions.

Notes

1. The size of individual classes is highly skewed; the smallest, Disposal of solid waste, accounts for 0,01% of patents and the largest, Basic Electric elements, for 6.49%.
2. This "bottom-up" approach selects 250 high technology product groups out of a 4 digit international trade classification.
3. The number of patents granted to Ireland, Portugal and Greece does not provide statistically significant results at this level of disaggregation. They have therefore been excluded from the analysis.
4. The correlations include data for the 116 subclasses with significant growth rates – all but Crystal growth and Disposal of solid waste.

CHAPTER 7

Sectoral Strengths and Weaknesses of Advanced Countries in Science

7.1. The Interest of Comparing Science and Technology

In the previous Chapters the sectoral specialization of technological activities of advanced countries is investigated by means of patent-based indicators. In this Chapter, the same approach will be followed in the analysis of the sectoral strengths and weaknesses of national scientific activities; the main indicators used to describe the countries' specialization will be publications and citations in the scientific literature. The interest of a comparison between the sectoral patterns of specialization in science and technology lies in the complex interrelations existing between the research activities of more fundamental nature taking place in institutions (universities, public and private research laboratories, etc.) devoted to the advancement of science on the one hand, and the research and development activities taking place in firms and other institutions (applied research laboratories, technical universities, etc.) focusing on technological innovation on the other.

Innovation is based upon a variety of sources, including the knowledge developed by the scientific community. We do not assume that technological knowledge is merely the pratical application of science, as stated for example by the "linear model"[1]. Without denying the importance of science as a source of technology, it should be remembered, as shown by several theoretical and empirical investigations[2], that innovation often stems from a number of different sources including the application or the diffusion of already existing knowledge.

An important aspect affecting the relationship between science and technology is the institutional setting of a country's research activities and the roles played by government bodies and business enterprises. Chapter 3 has already shown that advanced countries are characterized by specific combinations of institutional roles in funding and performing research activities. In some countries the role of firms is more relevant than in others, and the focus on the commercial applications of research may be greater, with more attention paid to research output which is proprietary in nature. The relationship between the public and the business sectors does not, however, depend on the aggregate flows of financial resources only, but also on the nature of the specific research projects undertaken.

The relationship between scientific and technological activities varies greatly across the fields. In some fields, a well established relation of scientific activities to practical applications can be found (e.g. chemistry and chemical products), while in other fields the same connection is lacking (it is, for example, difficult to relate mathematics to any specific technological sector). At the sectoral level, a great variety of links can be found between scientific and technological activities, but the relationship between fields of science and technology has not yet been systematically defined. A few studies have explored these links in detail in the case of very specific research specialties, where a clear relationship between scientific and technological activities could be established and where appropriate methodologies could be used (such as cross-citations among scientific publications and patents)[3].

While linking the sectors of specialization of scientific and technological activities of advanced countries is beyond the scope of this book, the analysis of sectoral patterns of specialization shown by science and technology can highlight the dynamics and the interaction of these two dimensions of research activities. In this Chapter, therefore, the sectoral strengths and weaknesses of advanced countries in science will be described along the same lines as the previous Chapters on technology in order to provide a term of reference for the more detailed analysis of technological specialization. An effort will be made to identify, in very general terms and at a highly aggregate level, the possible correspondence between the national specialization profiles emerging from scientific and technological indicators. Finally, the dynamics of scientific specialization shown by advanced countries over time and across output and impact indicators (scientific papers and citations) will be explored. In the next Chapter, further investigation will explore the degree of specialization shown by advanced countries in both technology and science.

7.2. The Sectoral Distribution of World Papers and Citations

Analysis of the scientific specialization of advanced countries is based on the number of publications and citations in the scientific literature drawn from the databases developed by CHI Research; the nature and structure of the databases, their relevance and limits have already been described in section 3.4. As we are here focusing on the sectoral distribution of national scientific activities, our analysis will be developed on the eight fields and 96 subfields in which all databases provided by CHI Research are disaggregated.[4] The eight fields include Clinical Medicine, Biomedical Research, Biology, Chemistry, Physics, Earth and Space, Engineering and Technology, Mathematics[5]. As in the aggregate analysis of Chapter 3, data for the number of scientific papers and the number of citations they receive from subsequent papers will be considered for three different periods: the 1973–78 and the 1979–84 periods are based on the 1973 journal set; the 1981–86 period is based on the new 1981 journal set (see section 3.4 for further details).

Differences in the sectoral patterns shown by papers and citations should be borne in mind; citation practices in particular differ strongly across fields. For example, in the 1973–78 data, the average number of citations received by papers in Biomedical Research was 14.02, while in Engineering and Technology it was only 2.61. In considering citation data, additional caution is required, as pointed out in previous Chapters. As the number of citations received by scientific papers builds up over time, if we group publications appearing over several years as we did in the three periods defined above, some distortion may be introduced as papers with differing probability of being cited by later papers are combined. This peculiarity of the data needs to be taken into account in analysing the results. However, citation data are grouped over a fairly long period and are used only to assess the impact of the scientific papers appearing over the same period.

After an overview of the sectoral structure of the database on scientific literature used, the following sections will provide a description of the fields of strength and weakness of each country examining the sectoral distribution and its change over time. A more specific indicator of the fields of relative specialization is provided by the index of Scientific Revealed Comparative Advantage (SRCA)[6], used in Chapters 4 and 5 in the analysis of patent data.

Table 7.1 shows the sectoral structure of the data drawn from the 1973 journal set. In the 1973–78 period, the distribution of world papers and citations in the 8 fields of science considered shows that Clinical Medicine accounted for 28.6% of all papers and 32% of citations and Biomedical Research for 15.6% of papers and 26.1% of citations. In terms of the proportion of world scientific literature, Chemistry follows with 16.7% of papers, Physics 13.4%, Engineering and Technology 9.8%, Biology 8.8%, Earth and Space 4.3% and Mathematics 2.9%. Earth and Space is the only field besides the biomedical sectors showing a share of citations (5.1%) greater than that of papers, while Physics has approximately the same share. In the second period, 1979–84, for which data from the 1973 journal set are available, the distribution across fields shows a growth in the share of papers in Clinical Medicine and a fall in that of citations, both being now at about 30%. Biomedical Research maintains its importance, with a quarter of all citations, and Biology and Chemistry also maintain their relative shares. Physics increases its share of citations to 15.2%, and Engineering has a slight reduction in its share of all papers[7].

The second database, from the 1981 journal set, for the 1981–86 period, shown in Table 7.2, reflects a substantial expansion of the number of journals and therefore of the papers and citations considered. The highest increase is in Clinical Medicine, which reaches 32.5% of all papers (and 33.1% of all citations), while Engineering and Technology (6.9%) and Chemistry (14.3%) have a smaller share of all papers and citations. Biomedical Research has a slightly higher share of papers (16.6%) and a greater share of citations (28.5%). Physics maintains its share of papers (13.4%), but its percentage of all citations is down to 14%.

These differences in the structure of the two journal sets have to be remembered when comparing the changing patterns of the various countries. The evolution of the journal set reflects the growing literature in Clinical Medicine and Biomedical Research, further expanding the weight of these fields (now close to 50%) in the database. In the citation pattern, the importance of these fields is even greater, affecting the overall structure of the citation activities.

TABLE 7.1.

WORLD PAPERS AND CITATIONS, 1973-78 and 1979-84

Percent distribution of scientific papers and citations, and average citations per paper in 8 fields of science

FIELD	1973-78 DATA PAPERS % of world	CITAT. % of world	Average cit. per paper	1979-84 DATA PAPERS % of world	CITAT. % of world	Average cit. per paper
CLIN. MEDICINE	28.63%	32.00%	9.34	30.12%	30.20%	3.11
BIOMEDICAL RES.	15.56%	26.11%	14.02	16.08%	25.82%	4.98
BIOLOGY	8.75%	5.76%	5.49	8.79%	5.70%	2.01
CHEMISTRY	16.70%	13.81%	6.91	16.08%	13.68%	2.64
PHYSICS	13.37%	13.29%	8.30	13.93%	15.19%	3.38
EARTH & SPACE	4.30%	5.06%	9.83	4.47%	5.48%	3.80
ENGIN. & TECHN.	9.76%	3.04%	2.61	8.16%	3.18%	1.21
MATHEMATICS	2.93%	0.93%	2.64	2.37%	0.74%	0.96
TOTAL	100.00%	100.00%	8.35	100.00%	100.00%	3.10

Source: CNR-ISRDS elaboration on CHI Research data

TABLE 7.2.

WORLD PAPERS AND CITATIONS, 1981-86

Percent distribution of scientific papers and citations and average citations per paper in 8 fields of science

FIELD	PAPERS % of world	CITAT. % of world	Average cit. per paper
CLINICAL MED.	32.50%	33.13%	3.59
BIOM. RESEARCH	16.58%	28.53%	6.06
BIOLOGY	9.26%	5.36%	2.04
CHEMISTRY	14.26%	11.32%	2.79
PHYSICS	13.40%	13.96%	3.67
EARTH & SPACE	4.68%	4.60%	3.46
ENGIN. & TECHN.	6.93%	2.42%	1.23
MATHEMATICS	2.40%	0.68%	1.00
TOTAL	100.00%	100.00%	3.52

Source: CNR-ISRDS elaboration on CHI Research data

7.3. The Patterns of Specialization of Advanced Countries by Fields of Science

Examination of national data for percentages and specialization indexes for publications and citations in 8 fields of science makes it possible to identify some national patterns. Table 7.3 shows the fields of specialization of the main OECD countries and the EEC as they result from the more recent database on papers and citations drawn from the 1981 journal set and appearing between 1981 and 1986. However, our analysis will also consider the results obtained from the previous data, for the 1973–78 and 1979–84 periods, based on the 1973 journal set[8].

In spite of the differences in the databases, a strong stability emerges in the specialization profiles of the advanced countries, as shown in Table 7.4 by the correlation coefficients between the SRCA indexes obtained for the same country in different databases. The correlation coefficients have been calculated between the vectors of the SRCA indexes for the number of papers in the three different databases considered, using the disaggregation in 96 subfields of science, in order to identify the consistency of national profiles of specialization in specific fields of research beyond the aggregate picture offered by the eight fields of science.

The first column of Table 7.4 lists the correlation coefficients between the specialization indexes of all countries for the two periods for which data from the 1973 journal set are available. They show that a very strong stability over time of national specialization profiles in science can be found for almost all countries, even in the highly disaggregated distribution of papers in 96 subfields of science. Eleven countries (including the US, Japan, the EEC, Germany and the UK) out of 19 have a correlation coefficient higher than 0.9; five more have a value above 0.8, while the lowest values are found for Spain and Portugal (0.58), countries which may well have expanded and partly changed their publication activity in international journals over the period considered.

Similar patterns of specialization are also found for many countries if we compare the national profiles resulting from different databases for different periods (i.e. indexes for the 1970s from the database described above, and indexes for the 1980s from the data drawn from the expanded 1981 journal set). While the US, Japan, Denmark, Ireland and Sweden have values above 0.8, values below 0.5 are found for Italy, Belgium, Portugal, Greece and Spain. For this latter group of countries, the change in the pattern of specialization may be due either to a change in sectoral distribution or to the different nature of the database considered. However, when the national specialization profiles resulting from the two journal sets are compared for an overlapping period (1979–84 for the 1973 journal set and 1981–86 for the 1981 journal set) the two pictures are again very close, with 13 countries showing a correlation coefficient greater than 0.8, including the US, Japan, and also the smaller non-English speaking countries listed above. The countries emerging with a more markedly different position in the two databases are Belgium (0.54) and the UK and Germany (0.69).

In summary, the importance of national patterns of "scientific accumulation" is borne out by this evidence as all countries show a fairly strong stability of their specialization in subfields of science. The two different databases used in this

TABLE 7.3.

SPECIALIZATION PROFILES OF ADVANCED COUNTRIES IN SCIENCE

Scientific literature: papers and citations in the 1981-86 database
Specialization indexes for 8 fields of science

Country		CLINICAL MEDICINE	BIO-MED. RESEARCH	BIOLOGY	CHEMIST.	PHYSICS	EARTH & SPACE	ENGINEER. & TECHN.	MATHEM.
United States	Papers	1.14	1.07	1.07	0.59	0.81	1.22	1.11	1.10
	Citat.	1.07	1.11	0.89	0.67	0.86	1.19	0.94	0.97
Japan	Papers	0.79	0.90	0.91	1.49	1.18	0.39	1.50	0.69
	Citat.	0.72	0.92	0.95	1.92	1.19	0.33	1.86	0.67
EEC	Papers	1.09	0.97	0.84	1.04	1.00	0.85	0.86	1.06
	Citat.	0.97	0.93	0.94	1.22	1.15	0.82	0.90	1.10
W. Germany	Papers	0.99	0.88	0.75	1.18	1.15	0.70	1.15	1.20
	Citat.	0.68	0.93	0.81	1.58	1.55	0.71	1.18	1.16
France	Papers	0.97	1.01	0.67	1.20	1.27	0.83	0.65	1.30
	Citat.	0.78	0.96	0.64	1.34	1.53	0.87	0.82	1.39
Un. Kingdom	Papers	1.21	0.99	1.11	0.75	0.73	1.00	0.98	0.87
	Citat.	1.17	0.96	1.25	0.89	0.70	0.93	0.90	0.98
Italy	Papers	1.13	0.87	0.49	1.28	1.18	0.86	0.65	0.90
	Citat.	0.99	0.65	0.44	1.71	1.46	0.90	0.76	0.95
Netherlands	Papers	1.14	1.11	0.95	0.84	0.96	0.95	0.60	0.91
	Citat.	1.02	0.99	1.02	1.09	1.01	0.83	0.65	0.92
Belgium	Papers	1.16	1.04	0.92	0.96	0.97	0.70	0.56	1.11
	Citat.	1.15	1.10	0.63	0.96	0.84	0.55	0.59	1.04
Denmark	Papers	1.72	0.90	0.65	0.44	0.72	0.66	0.34	0.77
	Citat.	1.41	0.79	0.78	0.56	1.15	0.49	0.46	1.11
Spain	Papers	0.60	1.18	0.73	2.38	0.94	0.46	0.46	0.98
	Citat.	0.58	0.89	1.09	2.54	1.19	0.33	0.77	0.99
Ireland	Papers	1.33	0.71	1.42	0.79	0.73	0.83	0.51	1.33
	Citat.	1.15	0.64	1.85	1.51	0.79	0.59	0.73	1.40
Portugal	Papers	0.73	1.03	0.58	1.07	1.48	0.77	1.40	2.25
	Citat.	0.79	0.92	0.48	1.26	1.55	0.60	2.06	2.02
Greece	Papers	0.38	0.66	2.20	1.47	0.78	3.77	0.01	3.07
	Citat.	0.51	0.30	4.93	2.50	0.51	2.01	0.04	5.04
Switzerland	Papers	1.26	0.96	0.46	0.88	1.23	0.61	0.79	0.64
	Citat.	0.84	1.09	0.35	1.10	1.59	0.51	0.75	0.50
Sweden	Papers	1.68	1.05	0.62	0.54	0.52	0.55	0.58	0.47
	Citat.	1.48	1.04	0.70	0.61	0.52	0.38	0.58	0.36
Austria	Papers	1.49	0.62	0.66	0.91	0.85	0.57	0.71	1.34
	Citat.	1.25	0.66	0.77	1.23	1.20	0.34	0.98	1.63
Canada	Papers	0.92	0.93	1.81	0.74	0.74	1.49	1.11	1.17
	Citat.	1.01	0.85	2.02	1.08	0.73	1.30	0.99	1.12
Australia	Papers	1.02	0.86	2.17	0.73	0.58	1.42	0.79	0.92
	Citat.	1.07	0.75	2.72	0.99	0.57	1.43	0.90	0.97

Source: CNR-ISRDS elaboration on CHI Research data

TABLE 7.4.

Correlation coefficients between specialization indexes (SRCA) for scientific papers resulting from different periods and databases

Database A1: 1973-78
Database A2: 1979-84
Database B: 1981-86

Countries	DATABASE A, 1973-78 with 1979-84	DATABASE A, 1973-79 with DATABASE B, 1981-86	DATABASE A, 1979-84 with DATABASE B, 1981-86
United States	0.96	0.82	0.86
Japan	0.97	0.87	0.87
EEC 12	0.93	0.07	0.79
W. Germany	0.91	0.56	0.69
France	0.87	0.63	0.81
United Kingdom	0.95	0.60	0.69
Italy	0.80	0.48	0.84
Netherlands	0.88	0.57	0.72
Belgium	0.82	0.41	0.54
Denmark	0.93	0.86	0.94
Spain	0.58	0.30	0.82 *
Ireland	0.99	0.90	0.92
Portugal	0.58	0.40	0.81
Greece	0.60	0.40	0.85
Switzerland	0.87	0.70	0.85
Sweden	0.94	0.87	0.93
Austria	0.93	0.67	0.76
Canada	0.96	0.79	0.81
Australia	0.92	0.75	0.80

Source: CNR-ISRDS, elaboration on CHI Research data

* Coefficient statistically significant to the level of 0.003%;
all the other coefficients are significant to the level of 0.001%

analysis lead to broadly similar results for most countries when the same period is considered and lead to more distant results when different periods are compared, in particular for countries where English has not yet become a commonly used language of scientific exchange. The generally high stability of the countries' profiles of relative advantages and disadvantages in 96 subfields of science ensures however that the databases considered are appropriate for the purposes of examining national specialization patterns in science.

After this qualification of the findings summarized in Table 7.3 for the eight major scientific fields, a description of the strengths and weaknesses of each country can be provided, based on the most recent results obtained from the 1981–86 data. An overview is offered by grouping the countries in three major clusters, defining a preliminary and very rough typology of the patterns of specialization in science.

The Bio-Medicine Cluster. At one extreme there are countries strongly specialized in Clinical Medicine and Biomedical Research. Sweden is the clearest case, with high indexes in these fields for both papers and citations and an average number of citations per paper well above the world average. In all other sectors Swedish science is less significantly represented. A group of countries combines a relative specialization in the Bio-Medicine sectors with other areas of strength. The United States has additional specializations in Earth and Space (the highest of all) and, for number of papers only, in Biology, Mathematics and Engineering. The UK has additional strengths in Biology. Belgium is also specialized in Chemistry and Physics. The Netherlands in Biology and Physics. Denmark in Physics and Mathematics (for citations only). The US and the UK show an average number of citations per paper in Bio-medical fields which is substantially higher than the world average.

The Sciences of Matter Cluster. The other extreme case is that of Japan, with a very high specialization for both papers and citations in Chemistry, Physics and Engineering, and a weak position in all other fields. The average citations received by the papers in these fields are, however, lower than the world average. A few European countries show a similar pattern of scientific specialization in Chemistry and Physics combined with other fields of strength. Germany comes close to this picture, but with a more even distribution across sectors. It has strengths in Chemistry, Physics, Mathematics and Engineering. In the first two fields, the citations received are higher than the world averages. France spreads its scientific strengths in Mathematics and Biomedical Research as well as in Chemistry and Physics, where its citation averages are at world levels. Switzerland also shows some specialization in Clinical Medicine and Biomedical Research, besides Chemistry and Physics, two fields with citation levels above the world averages.

Italy shows a strong performace in Chemistry and Physics, with citation values greater than the index from paper counts; the sectoral specialization index is greater than one also for publications in Clinical Medicine, while all other fields show substantial weaknesses. Only in Chemistry is the average number of citations received above the world average. Austria is somewhere in between the two clusters so far described, with a specialization in Chemistry and Physics (measured on citations), and strengths also in Clinical Medicine and Mathematics. The aggregate profile of specialization for the EEC shows strengths in Chemistry, Physics, Mathematics and Biomedical Research, and partly also for Clinical Medicine.

The Natural Sciences Outliers. Australia and Canada show a different pattern of specialization, with strengths in Biology, Earth and Space and Clinical Medicine; Canada has additional advantages in Mathematics and Chemistry (for citations only). These two countries emerge as having their own particular pattern of scientific specialization.

The Changes between the 1973–78 and the 1979–84 Data

The previous analysis of the correlations among the sectoral profiles of specialization emerging from different databases shown in Table 7.4 has documented the general stability of the national specialization patterns in science. It may be interesting however to examine the major changes in the data on scientific papers and citations for the 1973–78 and 1979–84 periods, drawn from the same 1973 journal set.

In the Bio-Medicine cluster, Sweden is in both periods the most extreme case, with high values for papers and citations. The US shows a stable pattern, with continuing strengths in these fields as well as in Earth and Space, Mathematics (although declining for citations) and Biology. The UK shows a fall over time of its specialization index in Biomedical Research, but increases its strength in Biology, reaches a new specialization in Earth and Space and keeps its advantage in Engineering. Belgium shows expanding strength in Clinical Medicine, a stable position in Chemistry, and new activity in Mathematics. Denmark shows similar strengths, but in Biomedical Research the index falls below one. Only the US and the UK have in both periods an average number of citations per paper higher than the world average.

In the cluster of sciences of matter, Japan shows exactly the same specialization profile, with a falling strength in Biology and growth in Engineering, the only field where the average citations received by Japanese papers are greater than the world average. The group of European countries with a similar pattern of specialization shows some changes. Germany shows a growing strength in Physics, where, as in Chemistry, the average citations are above the world's average. While Germany maintains a widely spread distribution of its scientific activities, its strength in Clinical Medicine shows some decline.

France shows a spread of its scientific strengths similar to the previous period, with a drop in specialization in Biomedical Research. In this period, French citations in Physics pass the world average mark. A stable pattern is also shown by the Netherlands, which loses strength in Biomedical Research, and gains specialization in Clinical Medicine, while Switzerland shows no change at all compared to the previous period. Both countries have citation averages above the world averages in almost all fields. Italy maintains its strong performace in Chemistry and Physics, with greater specialization indexes for citations than for papers; a decline in Biomedical Research is paralleled by growing strength in Clinical Medicine, but in no field does the average number of citations received come close to the world average. The EEC aggregate appears stronger in Chemistry, Physics and Mathematics, and slightly weaker in the Bio-Medical fields. Austria maintains its mixed specializations, with a growth in Mathematics and a slight drop in Chemistry.

In the natural sciences outliers, Australia and Canada again show their particular pattern of specialization, with growing strengths in Clinical Medicine, a drop in Mathematics for Australia, and a stable profile for Canada.

The patterns of scientific specialization in advanced countries have led to the identification of two different "models" of scientific activity: the first is based on the Bio-Medicine sciences, and can be found in the US, the UK and other smaller European countries, from Sweden to Denmark, Belgium and the Netherlands. The second one is based on the sciences of matter, Chemistry and Physics in particular, which characterizes most European countries and Japan.

Summing up the evidence so far described, the growing importance of Bio-Medical sciences is common to most countries, and it is reflected also in the changes of the 1981 database. In terms of relative patterns of specialization, attention should be drawn to the rather stable picture shown by the national profiles from the early 1970s to the late 1980s. This lends support to the hypothesis of the importance of "scientific accumulation" in research activity, suggesting a pattern of cumulative knowledge which parallels the one already found in the analysis of technological specialization in Chapter 4. A more detailed study of the degree of specialization will be developed in the next Chapter, relating it to the size of a country's scientific activities.

7.4. Comparing Scientific and Technological Specialization

A comparison is now possible between the specialization profiles of the advanced countries drawn by patent indicators for technology and by bibliometric indicators for science. This can help to identify a few broad regularities in national research efforts and the specific characteristics of national systems of innovation. A large basis of scientific research and dynamic technological activity are both important aspects of a country's innovative system; they draw upon each other's efforts, stimulate original work and feed back on one another, leading to advances which would not be possible without the parallel development of science and technology. At the sectoral level, however, no clear correspondence exists between the breakdown by fields used in this Chapter for scientific papers and that employed above for technological activity. Some science fields, such as Mathematics, lead to basic knowledge which is important across the whole spectrum of applied research. The results of other science fields, such as some areas of Physics and Engineering, may prove relevant to a wide range of technological activities, especially in mechanical and electrical-electronic sectors.

Closer links between scientific research and technological activities can be found at a more disaggregated level if we look at the available disaggregation in 96 subfields of science and at individual patent classes. Areas where some correspondence can be found include Agriculture and food, Natural resources (i.e. some subfields of Earth sciences and the patent class of Petroleum and natural gas), various Chemical fields (Organic, Inorganic, Polymers, etc.), Pharmaceuticals, some electronic-related areas (Computer science and Solid state physics with the patent classes of Office computing and Electronic components), and the field of Aerospace. As the SIC patent classification employed is based on product classes and is largely related to industrial activities, a number of scientific fields which are associated with service activities, such as Medicine and Bio-medical research, cannot be matched to the available technological fields; only a minor part of this research is reflected in the manufacturing of drugs.

In the fields where a closer link may be found between science and technology (although the areas should be identified in far greater detail) we can indeed find that technological innovation draws upon scientific research and viceversa, but nothing leads us to expect that countries active in such fields will show a parallel strength in science and technology. In other words, no clear correspondence can be found between the specialization profiles previously described for science and technology, due to the differences in the the nature of the activities and of the indicators considered. Comparison of the specialization profiles of the advanced countries in science (as measured by bibliometric indicators) and technology (as described by patenting in the US, see Chapter 5) may however be of some interest in highlighting regularities and national specificities in the pattern of science and technology.

Countries such as Japan, Germany, France, Italy, Switzerland and Spain, whose scientific activities show a relative specialization in the "sciences of matter", with strengths in Physics, Chemistry and (in some cases) Engineering, tend to concentrate their technological specializations in chemical and mechanical fields (in classes such as Specialized industrial machinery, Motor vehicles, and Railroad equipment). Only Sweden combines an advantage in these technological sectors with a contrasting pattern of scientific specialization in Bio-Medicine sciences.

A more specific correspondence for the European countries specialized in "sciences of matter" (Germany, France, Italy and Switzerland) is found between scientific excellence in chemistry and technological advantage in some chemical-related patent classes, including Organic chemicals, Agricultural chemicals, Plastic and synthetic materials, Soaps and detergents, Drugs and medicines. However, the UK, Belgium and the Netherlands, which show no strong scientific specialization in chemistry, also include some chemical-related patent classes among their technological strengths.

Japan, on the contrary, combines its specialization in the "sciences of matter" with strong technological specialization in electrical and electronic fields, as does the Netherlands. If we look at the 96 subfields of science, we find that Japan is strongly specialized in all periods in Applied physics, General physics, Optics and Electrical and electronic engineering. The Netherlands, on the other hand, is specialized in Chemical physics, Acoustics and General physics, with other varying areas of specialization in individual periods, including Computers. In both cases the linkage between scientific and technological activities is quite clear. For the countries specialized in the Bio-Medicine sciences (Sweden, the US, the UK, Belgium, the Netherlands and Denmark) no clear correspondence with product-based technological strengths is possible. Sweden, as already pointed out, has technological strengths in mechanical and resource-based sectors. The Netherlands shows technological specialization in electric and agriculture-related fields. However, the UK, Belgium and Denmark show a link between their scientific specialization in Bio-Medicine, and in particular their strength in the subfield of Pharmacology, and their technological advantage in the patent class Drugs and medicines.

In the case of the United States, it is pointed out in Chapters 4 and 5 that its specialization profile in technology needs to be investigated by using its patenting activity abroad. Examination of the areas of greatest US specialization resulting from US patents registered at the European Patent Office, classified by IPC classes (based on human needs rather than products, see Chapter 4), reveals that Medical preparations and Bio-chemistry are two of the top five classes of greatest US specialization. Health is also a class of relative US advantage. The strong scientific specialization of the US in Bio-Medical fields is therefore paralleled by the evidence from US patenting abroad.

In other sectors, the US shows a parallel strength in science and technology in the areas of Food (the science subfields of Agriculture and food sciences and Dairy and animal sciences, and the IPC technological class of Foodstuffs, again for patents registered at the EPO), Natural resources (some subfields of Earth sciences and the technology class of Mining), Electronics (science subfields of Computers and Electrical and electronic engineering, and technological classes of Computing and Information instruments) and Nuclear research (the science subfield of Nuclear technology and the patent class of Nuclear physics).

Most of the evidence presented above for both science and technology indicators is consistent in all the available data from the mid–1970s to the late 1980s. However some changes over time should also be pointed out. The lack of clear technological specialization in medical-related patent classes shown by countries such as the Netherlands, Belgium and Denmark, for instance, may also be related to the more recent emergence of their scientific specialization in Bio-Medicine sciences; one may suggest that broader and more consolidated scientific effort is required to translate a relative advantage in science into a relative strength in medical-related technologies.

While the regularities described here are very far from identifying a clear pattern of relationships between sectoral specializations in science and technology, a few insights have been obtained into the fields where parallel strengths (or weaknesses) tend to emerge for advanced countries. A fuller picture of the links between science and technology systems has however to consider in greater detail the nature of national systems of innovation, the balance between basic research, applied research and development efforts, the presence and role of institutions (universities, public and private research institutes, business sector, etc.), the funding patterns and the sectoral priorities identified in each country.

7.5. The Sectoral Distribution of the Resources Devoted to Science

An example of the kind of information which could be usefully integrated in this analysis is a comparison of the funding for academic research in different fields in the US, Japan, Germany, France and the UK, developed at the Science Policy Research Unit at the University of Sussex (Martin, Irvin and Isard (1991), US National Science Board (1989, p. 99)). This study examines the distribution of funds for academic and related research, which is a major input for scientific activity, in the fields of Life sciences (including all Bio-Medical fields and Agricultural science), Physical sciences (Chemistry, Physics, Astronomy, etc.), Environmental sciences (Geology, Atmospheric science, Oceanography, etc.), Mathematics and computer sciences, Engineering and other non-science fields.

The results show that the US devotes almost 50% of its academic funds to Life sciences, while the other countries have a share between 30 and 35%. Physical sciences are the second largest destination of academic funds, but here the US (with Japan) is at the bottom of the list, while France and Germany spend 25 to 30% of their resources in this field. Japan, on the other hand, has an above average share in Engineering (22%, with the other countries between 10 and 15%). The UK has the highest shares (above 5%) in the two remaining fields, Environmental sciences and Mathematics and computer science, with the other countries closely behind (see US National Science Board, 1989, p.99, 289).

This evidence parallels some results of the analysis of scientific papers, which show clear US strength in all Bio-Medical fields, in Engineering and in some Environmental sciences. The large proportion of Japanese funds devoted to Engineering corresponds to the marked advantage found for scientific papers in this field. However, Japan also has a scientific strength in Chemistry and Physics in spite of the lower proportion of its funds devoted to academic research in Physical sciences. Less clear-cut is the evidence for European countries, where the fields of apparent priority in research funding often fail to correspond to the areas of relative strength measured by scientific papers, suggesting that a much more detailed analysis would be required here, including also aspects such as the institutional setting and the particular forms of research funding. This example shows the interest and the complexity of a parallel investigation of different aspects of the science and technology systems of advanced countries. Some links can be identified between the fields of relative national strength resulting from different indicators of national research activity, but only a more thorough investigation of the national system of innovation can provide an adequate account of the complex relationship between science and technology activities.

While it may be tempting to look upon the success of some countries in particular fields as a model for national policies in science and technology, this evidence suggests that the particular combinations of a country's relative strengths are more the outcome of long-term activities and institutional settings than a short-term result which could be replicated elsewhere. Nevertheless, some lessons for policy can be drawn from the pattern of linkages between some selected fields of greater specialization in science and technology pointed out in this Chapter. The next Chapter will move from the sectoral analysis of the areas of national specialization to a more general view of the extent to which countries concentrate their research efforts and innovative activities in a few fields, or spread them across many areas, again comparing science and technology indicators.

Notes

1. According to the linear model, an innovation is the result of several steps starting from the invention (which is assumed to embody a consistent part of scientific knowledge), and going on to the innovation (i.e. the application of the scientific knowledge) and the diffusion (its market impact) (for a critique, see Kline-Rosenberg, 1986).
2. An overview of the discussion on the relationship between science and technology can be found in Price (1984, 1986), Rosenberg (1982), Chapter 7; Archibugi (1986).
3. Among the more recent studies, see Narin and Noma (1985), Coward and Franklin (1989), Van Vianen et al. (1990).
4. The original database provides 106 subfields. In our analysis a few subfields with a very small number of publications and citations have been merged.
5. Psycology was previously included in these fields, but was later dropped and transfered to the Social Science Citation Index.
6. The index is equal to the ratio between a country's share of publications (or citations) in the scientific field j and the total share of publications (or citations) of the country in the world total.
7. Full information on the number of journals included in the databases and on the papers and citations considered is provided in Chapter 3. The rates of change of the number of papers in the 8 fields of science show the changing composition of the database, but are not a reliable

indicator of the changes in international scientific publishing activity, as they refer only to the fixed journal set database. Clinical Medicine is the area showing the fastest growth, 8% between the first and the second period, with the number of papers in Biomedical Research increasing by 6%; Physics and Earth and Space also show expansion in scientific activity, with 7% growth rates. For Engineering and Mathematics on the other hand, the number of papers falls by 14% and by 17% over the two periods.

8. Full information for the 8 fields and the 96 subfields of science, drawn from the two databases (1973 journal set and 1981 journal set) is included in a preliminary research report; see Pianta and Simonetti (1990).

CHAPTER 8

Degree of Specialization and Size of National Scientific and Technological Activities

8.1. The Analysis and Measurement of the Degree of Specialization

In the previous four Chapters the specialization profiles of advanced countries have been identified according to both technology and science indicators. The mapping of sectoral strengths and weaknesses, however, also requires a description of how concentrated (or spread out) national science and technology activities are across fields. Beyond analysis at the sectoral level, in fact, the resulting aggregate tendencies, and differences existing among countries, should be explored. This Chapter examines the extent to which countries focus their efforts in a few areas of science and technology, or distribute their resources over many different fields. This analysis qualifies the study of national sectoral specialization developed in the previous Chapters, and constitutes an original contribution of this report. In particular, the following questions will be addressed:

i) How can we identify the overall degree of specialization[1] shown by individual countries, and which indicator can be used for this purpose?

ii) If, as we would expect, significant differences are to be found across countries, what are the key factors affecting the national degrees of specialization, and what can we learn from this as regards the nature of national specialization profiles? In particular, what is the influence of the size of a country on its degree of specialization?

iii) Is there a trend over time in the degree of sectoral specialization in S&T for advanced countries? In other words, are countries increasing or decreasing the dispersion of their S&T activities across sectors?

iv) Do the trends on the degree of specialization follow the same dynamics in science and technology, or is there any divergence?

v) How are the national degrees of sectoral specialization related to the aggregate trends on S&T activities outlined in Chapter 3?

Some of the above questions have already been tackled in a different context and for different purposes, by economists in the field of international trade. The specialization in trade and production has often been examined in cross-country studies, the results of which have pointed out that small countries have a higher degree of openess to international trade and that higher levels of specialization in production are imposed upon them as the size of their domestic productive capacity does not allow them to cover all industries uniformly. In more general terms, an inverse relationship has been shown between the size of each country and its trade specialization.[2]

Analyses over time have provided evidence on two related phenomena. On the one hand, an increasing productive specialization has been found for many countries and sectors (a comprehensive survey is provided in Porter, 1990). On the other, this has been combined with a growing convergence among industrial countries in terms of several economic indicators, including production, productivity, and investment (see Dollar and Wolff (1988, 1989), Baumol et al. (1989)). A similar process of convergence is identified in Chapter 3 for the aggregate indicators of resources devoted to technology in advanced countries. The following sections of this Chapter will investigate whether growing specialization is also taking place in the S&T system and how it is related to the convergence in the aggregate efforts in S&T. The existence of a parallel relationship between the size of national S&T activities and the degree of specialization will be explored.

In this Chapter, we will firstly consider patterns in technology, using our empirical evidence on patents. Secondly we will examine the evidence from scientific activity, by using bibliometric indicators. The central methodological and empirical issue of this Chapter is the measurement of the degree of specialization shown by national sectoral distribution of S&T activities across different fields. A number of indicators have been considered in order to provide a satisfactory description of the patterns of specialization of individual countries.[3] The best indicator we have identified is the chi square value, calculated by comparing the percentage distribution across sectors of the science or technology activities for a given country and the percentage distribution of the world total. The formula used is the following:

$$\chi_i^2 = \Sigma_j[(p_{ij} - p_{wj})^2/p_{wj}]$$

where:

i is the country considered
j indicates the sectors of the distribution
p_{ij} is the percentage of the variable considered held by country i in class j
p_{wj} is the percentage of the variable considered for the world total in class j.

The vectors containing the percentage distribution of activities (i.e. patents, patent citations, scientific papers and paper citations) across fields for each country in the databases considered have been compared to the vector of the sectoral percentage distribution of the world shown by the same databases. Chi square values were calculated for each column of a matrix where each cell contains the share of patents (or patent citations, scientific papers, and paper citations) in the various sectors for a given country. This provides a measure of how national profiles differ from the world sectoral profile, which can be used as a simple index of the degree of specialization of each country.

If a country has the same percentage distribution as the world, its chi square value will be equal to zero. In other words, the country would not show any sectoral specialization.[4] The more a country presents marked sectoral strengths and weaknesses, the greater the value of the chi square. Since the chi square values are calculated on the countries' percentage distribution of activities across sectors, they are not influenced by the number of observations available for each country. Use of the percent distribution across sectors solves the problem of the different number of patents or scientific papers of individual countries, but the effect of the country's size on their sectoral distribution still has to be examined (see sections 8.3 and 8.4)

The use of chi square values as an index of the degree of specialization makes it possible to develop a preliminary analysis in the following fields: i) to compare the degree of specialization of different countries; ii) to compare their degrees of specialization measured on different databases; iii) to examine the changes over time in the position of individual countries; iv) to compare degrees of specialization in science and technology. More detailed analysis of the degree of specialization will also be required to consider the effect of the different size of countries, and will be developed in Chapter 10.

8.2. The Degree of Specialization of Technological Activities

Table 8.1 shows for all countries the chi squares values measured on patents registered in the US and at the European Patent Office according to the IPC classes (described in Chapter 4) and on patents granted and patent citations in the US according to the SIC classes (described in Chapter 5).[5] Given the less differentiated nature of domestic patenting activity, pointed out in Chapter 4, it is no surprise that the US shows the lowest degree of specialization in the data based on US patents, and that the EEC aggregate has the lowest specialization at the EPO, the patent institution which is the main vehicle of appropriability within the EEC internal market.

West Germany, the UK and France follow with the lowest degrees of specialization at the EPO. The same countries are found, but in different order, when looking at patenting in the US, according to both the IPC and SIC classifications. Canada, Italy, the Netherlands, Denmark, Belgium, Sweden, Switzerland and Spain (generally in this order) follow with growing degrees of specialization measured by the chi square values. Less reliable values are found for the three smallest patenting countries, Ireland, Portugal and Greece. For the majority of countries, the degree of specialization appears fairly consistent in all the databases considered. For patents granted in the US in the 1980s, even comparing two very different sectoral classifications such as the SIC and the IPC (columns 2 and 5 of Table 8.1) the ranking of countries is very similar, with a rank correlation coefficient of 0.96. The most significant differences emerge when comparing the degrees of specialization measured on patents in the US and on patents at the EPO, where the rank correlation coefficient is 0.84 when data are based on the same IPC classification (columns 6 and 5), and 0.83 when data are based on different classifications (the IPC for the EPO and the SIC for the US; columns 6 and 2). These results show that, as expected, differences in the Patent Office database used are more important than the differences in the types of sectoral classifications employed. The degree of spe-

TABLE 8.1.

THE DEGREE OF TECHNOLOGICAL SPECIALIZATION - CHI SQUARE VALUES
Chi square values of the percent distributions by sectors of patents and patent citations for advanced countries:
A. Patents and citations in the US by 41 SIC classes, 1975-81 and 1982-88
B. Patents granted in the US 1981-87 and patent applications at the EPO 1982-87 by 31 IPC classes

Countries	A. Chi squares by 41 SIC classes				B. Chi squares by 31 IPC classes	
	Pat. gr. in the US 1975-81 (1)	Pat. gr. in the US 1982-88 (2)	Pat. cit. in the US 1975-81 (3)	Pat cit. in the US 1982-88 (4)	Pat. gr. in the US 1981-87 (5)	Pat. appl. at the EPO 1982-87 (6)
US	0.94	1.31	1.05	2.06	1.61	7.84
Japan	13.46	14.68	12.96	14.96	20.98	19.23
EEC	3.84	4.50	5.76	6.90	4.49	3.23
W.Germany	8.16	10.05	13.51	15.39	9.39	3.70
France	4.00	3.86	4.01	3.83	8.46	10.89
Un.Kingdom	5.91	6.85	10.43	17.91	5.97	5.27
Italy	21.85	24.53	25.55	25.21	26.92	33.62
Netherland	23.06	20.46	27.52	22.48	21.72	23.35
Belgium	30.72	38.84	56.02	110.56	28.89	38.01
Denmark	24.63	31.88	41.06	62.40	n.a.	n.a.
Spain	46.88	53.52	88.73	101.09	n.a.	n.a.
Ireland	77.99	22.42	84.78	50.58	n.a.	n.a.
Portugal	139.81	212.25	289.36	299.57	n.a.	n.a.
Greece	96.13	89.96	153.46	290.15	n.a.	n.a.
Canada	12.38	14.09	16.56	13.41	18.63	n.a.
Switzerland	36.16	34.39	39.54	56.12	41.41	24.95
Sweden	24.72	24.74	23.70	23.15	32.80	42.81

Source: CNR-ISRDS elaboration on CHI Research and EPO data
n.a. - not available
NOTE: The EEC data by IPC classes include only W. Germany, France, Italy, Netherlands, Belgium, U.Kingdom.

cialization is generally higher when measured on patent citations than on patents granted; the impact of countries' technological inventions appears less uniformly distributed across sectors than the number of patents they register.

If we examine changes over time in patenting in the US, these indicators of the degree of technological specialization show a general increase in the values for both patents and citations. Only France and the Netherlands show a drop in their degree of specialization, while Canada, Sweden and Italy show rising specialization for patents and a (moderate) drop for citations. The patterns of the three smallest countries are again less clear due to the little number of patents registered. For most countries, the increase over time of the degree of specialization is much higher for patent citations than for patent counts, suggesting a more rapidly growing differentiation in the technological activities with greater impact. This result is, however, likely to be affected by the uneven distribution of patent citations across classes, and by the shorter time period for which citations to recent patents are available.

8.3. The Relationship between Size and Technological Specialization

A key issue in the exploration of the dynamics of technological specialization of the advanced countries is the analysis of the relationship between the size of the technology base and the degree of specialization. The existence of regularities in this relationship can highlight the possible "paths of specialization" followed by countries as they expand their S&T activities and search for technology based competitive advantages in international markets. We expect that as the size of a country grows, its science and technology activities are spread more evenly across sectors. In terms of the indicator of specialization so far used, we have already seen in Table 8.1 that smaller countries generally have higher degrees of specialization and large countries lower values.

The results of the previous section make a cross-country study of this relationship possible, allowing to identify the position of individual countries in relation to the overall distribution. As an indicator of the level of a country's specialization we will use the chi square values shown in Table 8.1, and as an indicator of the technology base we will use the cumulative R&D expenditure for the same periods, 1975–81 and 1982–88[6]. Cumulative R&D has often been used as an appropriate indicator of the amount of national technological activities; it avoids the random fluctuations of annual data and stresses the cumulative nature of technological efforts. Furthermore, it is closely related to other more general indicators of country size, such as GDP.

We have plotted the position of each country along these two variables in figs. 8.1, 8.2 and 8.3; for ease of presentation the two axes have been transformed into logarithmic scale, and the resulting relationship is therefore a linear one.[7] The regression lines are drawn in order to show the general pattern of distribution and to identify the countries whose degree of specialization, in relation to their size, is above or below the general trend. Obviously, a country's relative position depends on the set of countries considered, which is in our case the fairly homogeneous group of more advanced OECD countries. The variety of the databases considered allows us to assess the stability of distribution: 1) across two different classifications, 2) across two different patent institutions, 3) over time, 4) between indicators of simple count (the number of patents) and of impact (citations).

As expected, all the grahps show a consistent inverse relationship between the size of the technology base and the degree of specialization. While we have already discussed in the previous section the international differences in the absolute levels of specialization, here we can compare the position of individual countries, and their shifts, in relation to the volume of their S&T activities. Figure 8.1 shows, for the period 1981–87, the patterns of specialization emerging from US and EPO patent counts, disaggregated by IPC classes. A clear inverse relationship emerges. The distribution of the two sets of data is fairly similar, with the notable exception of countries where the "domestic market effect" (see Chapter 4) emerges (the US has a low specialization degree in the domestic market, and the EEC and Germany have a similarly low index at the EPO). Japan has a considerably higher specialization degree than would be expected from the size of its S&T activities. Italy and, to a lesser extent, Sweden also have rather high specialization levels, while Great

Degree of specialization and size of technological activity

Chi square values of the distribution by IPC classes of patents in the US and at the EPO, and cumulative R&D expenditure, 1982-88
1 - US: Patents granted in the US, 1981-87
2 - EPO: Patent applications at the EPO, 1982-87

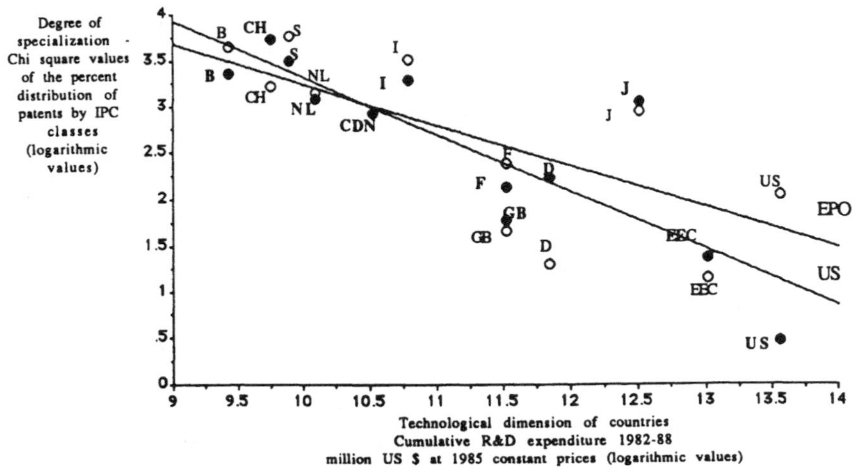

○ normal = calculated on patents granted at the EPO, 1982-88
● bold= calculated on patent applications in the US 1981-87

Legend

B= Belgium	NL= Netherlands
CDN= Canada	S= Sweden
F= France	CH= Switzerland
D= W. Germany	GB= United Kingdom
I= Italy	US= United States
J= Japan	EEC= European Community

Fig. 8.1.

Britain and France appear to spread their technological activities across a broader range of sectors.

Figure 8.2 shows the same relationship for the degree of specialization measured on patents granted in the US according to the SIC classes for two periods, 1975–81 and 1982–88. Over time, a general upward shift is clearly visible. The countries' relative positions are confirmed, with the US, Great Britain and France showing degrees of specialization below expected, while Japan, Italy, Switzerland and Spain present higher levels of specialization. Figure 8.3 presents data on patent citations in the US for the same periods. The upward shift of the regression line from the first to the second period is even more evident than for patent counts. For countries such as the US, Great Britain and Belgium, the specialization degree increases sharply, while France, Canada and the Netherlands are the only countries showing a slight reduction.

The increase over time in the degree of specialization shown by most countries

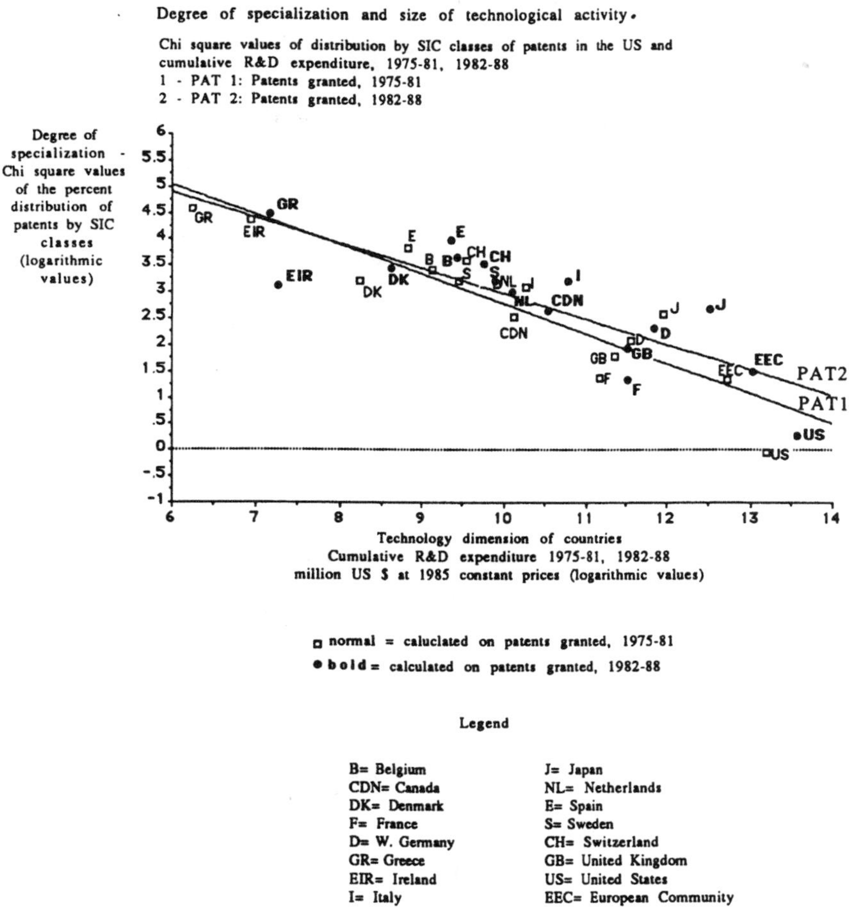

Fig. 8.2.

is particularly significant because it goes against the expected changes. Over time, all countries have increased the volume of their patents, and this growth could lead to a more even distribution across sectors. On the contrary, an increased distance is nearly always found between the countries' profiles of activity and the world sectoral distribution of patents. In the following sections the degree of specialization of national scientific activities will be considered, and a comparison attempted between degrees of specialization in science and technology.

8.4. The Degree of Specialization of Scientific Activities

A parallel investigation to the one developed so far on the degree of technological specialization is possible for scientific activity, documented by the bibliometric indicators examined in Chapter 7. In order to produce accurate information on the degree of specialization in science, data on the number of scientific papers and of

Degree of specialization and size of technological activity.

Chi square values of distribution by SIC classes of patent citations in the US and cumulative R&D expenditure, 1975-81, 1982-88
1 - CIT1: Citations 1975-84
2 - CIT2: Citations 1982-88

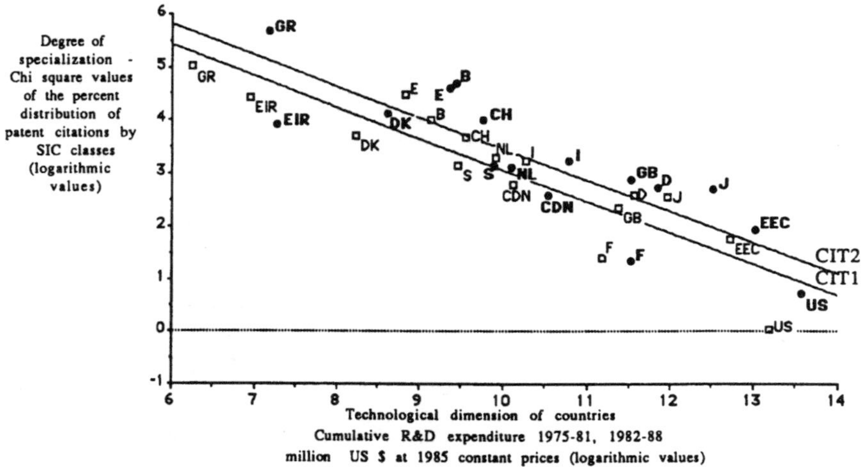

□ normal = calculated on patent citations, 1975-81
● **bold** = calculated on patent citations, 1982-88

Legend

B= Belgium	J= Japan
CDN= Canada	NL= Netherlands
DK= Denmark	E= Spain
F= France	S= Sweden
D= W. Germany	CH= Switzerland
GR= Greece	GB= Great Britain
EIR= Ireland	US= United States
I= Italy	EEC= European Community

Fig. 8.3.

the citations they receive from other publications, drawn from the CHI Research databases described in section 3.5 and in the previous Chapter, will be analyzed here at a greater level of disaggregation. For all countries we now consider the number of papers and citations in 96 subfields of science into which the 8 fields examined in Chapter 7 are subdivided.[8]

The analysis of the degree of specialization will be developed on both the databases available. The first is drawn from the journal set defined in 1973, and divided into two periods, 1973–78 and 1979–84. The second database, drawn from the new and expanded journal set defined in 1981, is divided here into two subperiods, 1981–83 and 1984–86, in order to obtain information on the changes over time in national degrees of scientific specialization. It should also be remembered that the national specialization profiles resulting from the different databases are strongly and positively correlated.

The methodology followed for the analysis of the national degrees of specializa-

TABLE 8.2.

THE DEGREE OF SCIENTIFIC SPECIALIZATION - Chi square values

Chi square values of the percent distribution by subfield of papers and paper citations for advanced countries
A. 1973-84 database: 1973-78 and 1979-84
B. 1981-86 database: 1981-83 and 1984-86

Countries	A. Data from 1973-84 database papers 1973-78	citations 1973-78	papers 1979-84	citations 1979-84	B. Data from the 1981-86 database papers 1981-83	citations 1981-83	papers 1984-86	citations 1984-86
USA	10.26	3.15	9.47	3.76	8.27	4.03	7.03	4.22
Japan	54.80	61.11	40.45	53.11	38.86	50.75	36.34	48.62
EEC	5.37	5.11	5.08	5.19	3.82	4.21	3.54	4.38
W. Germany	20.29	29.68	18.44	33.05	13.76	21.92	13.23	25.21
France	34.73	24.69	20.92	17.79	15.04	18.04	14.62	15.37
Un. Kingdom	16.41	11.78	16.84	13.78	12.00	10.29	11.97	10.75
Italy	52.38	43.06	38.12	37.66	33.71	38.68	30.92	34.88
Netherlands	19.20	19.92	19.76	18.37	14.20	14.10	14.26	20.02
Belgium	23.97	19.94	23.11	23.62	37.69	22.27	23.59	23.94
Denmark	73.24	50.51	75.88	51.42	61.50	39.43	71.75	46.05
Spain	107.30	50.55	64.93	51.08	90.31	72.93	75.20	83
Ireland	160.54	116.66	120.53	140.61	121.26	131.00	73.03	81.08
Portugal	93.54	140.89	68.62	114.17	90.85	326.00	77.23	157.09
Greece	64.37	94.30	50.80	75.21	63.61	107.22	54.59	91.5
Canada	25.59	27.39	30.85	26.82	18.74	22.23	18.99	22.27
Switzerland	41.33	35.57	30.90	32.11	38.14	46.03	30.61	46.39
Sweden	57.71	41.26	70.25	40.58	63.50	41.95	56.15	47.17

Source: CNR- ISRDS elaboration on CHI Research data

tion in science is the same as that followed in the previous sections of this Chapter for the study of technological specialization based on patent data. Chi square values were calculated for each country in order to measure how national patterns of specialization in scientific subfields differ from the aggregate distribution of world papers and citations. Chi square values were calculated for the two subperiods of the two databases, for both papers and citations, and are shown in Table 8.2. They allow an assessment of the degree of specialization across countries and over time as well as a comparison of the dynamics of paper counts and citations.

The EEC aggregate presents the lowest chi square values measured for the number of papers in all periods, while the US shows the lowest degree of specialization measured for paper citations in all periods. While the EEC spreads its scientific publications across subfields more evenly than the US, the EEC papers with the greatest impact, as shown by the citations they receive, are not distributed as widely as those of the US. However, this result may be due to the overrepresentation of US (and generally English language) journals in the database considered. In fact, a paper published in English has higher probability of being cited than those in other languages.

The UK is the country showing the next lowest degree of specialization for both papers and citations across subfields of science, and the importance of English language journals is evident also in this result. Germany and France follow, with the former showing a lower degree of specialization in papers, and the latter presenting lower chi square values for paper citations in all periods. The Netherlands, Canada and Belgium have a more irregular pattern in the different periods, in a few cases showing even lower chi square values than Germany or France.

Among the large countries, Japan presents one of the highest degrees of specialization. In Chapter 3 we have already found that Japan has a low amount of scientific production, compared to both its R&D inputs and its patent output; in addition, Chapter 7 has shown that the Japanese scientific community concentrates its energies in fields closely related to technological knowledge. In other words, the Japanese innovation system not only focuses more on technological than on scientific activities, but within science it also concentrates efforts in areas more closely linked to technology. As a result, its degree of specialization in science is substantially higher than the pattern shown by the set of countries considered.

Italy, Switzerland, Sweden and Denmark have a high degree of specialization, concentrating their scientific efforts in selected subfields. Even higher, but also less reliable, are the chi square values shown by Spain, Greece, Portugal and Ireland. In general, the ranking of countries according to the degree of scientific specialization appears less stable than that found for the level of technological specialization measured on patent data. More marked differences emerge between national positions in papers and citation data. Different patterns are found on comparing the chi square values for paper counts and citations. In the first period, 1973–78, the degree of specialization shown by all papers is generally higher than for citations; among the largest countries, only Japan, Germany and the Netherlands contrast with this pattern. In other words, in most countries "good" papers are more evenly distributed across fields of science than all papers. In the 1979–84 period, Japan and Germany maintain their greater specialization in citations than in paper counts, and are joined by Switzerland and Belgium.

A different picture emerges from the new database built on the 1981 journal set. In the period 1981–83 France, Italy and Canada join the countries with a greater degree of specialization in citations than in papers. In the last period, 1984–86, only the US, the UK and Sweden show a more even distribution across subfields for citations than for papers. Particular caution is required in interpreting this result. Firstly, we should remember the higher citation rates and the more even distribution of citations across sectors shown by English-speaking countries due to the nature of the database. Secondly, citation data build up over the years, and a much larger number of citations is available for earlier papers than for later ones, which may favour a more even distribution across scientific subfields shown by the data for the earlier periods.

If we now turn our attention to the changes over time in the degree of scientific specialization, the pattern shown by paper counts is clearly falling for almost all countries. In the first database, comparison of the two periods 1973–78 and 1979–84, shows that France, Switzerland, Italy and Japan are the main OECD countries spreading their scientific activities more uniformly across fields, and thus showing lower chi square values for both paper counts and paper citations. Several other countries show the same falling specialization over time for paper counts only (the US, Germany and Belgium), or for citations only (the Netherlands, Canada and Sweden). The UK and Denmark alone show an opposite pattern of growing specialization over time in both paper counts and citations.

If we consider the second database, the fall in the degree of specialization between data for 1981–83 and 1984–86 shown by the number of scientific papers is even more evident, and only the Netherlands, Denmark and Canada concentrate their scientific activities in fewer sectors. Citation data, on the other hand, show an increasing degree of sectoral specialization which, as already pointed out, is affected by the short time period considered.

The patterns of change in the national degrees of specialization shown by scientific papers suggest that over time countries devote additional resources to expanding their scientific activities in the areas of their greater weakness, resulting in a distribution of the specialization indexes across subfields of science which is closer to that resulting from all world papers. This more even sectoral distribution of national activity is an expected outcome of the general increase in the resources, human and financial, devoted to scientific research in most countries. As the countries' activity increases, we can expect a pattern of scientific output closer to the world distribution to emerge.

This pattern of decreasing specialization of national scientific activity is rooted in the nature of scientific activity: the non-proprietary nature of scientific knowledge and the availability of state of the art knowledge published by international scientific journals make it possible to reach or remain at the world level, at the frontier of scientific knowledge in a variety of areas, without any need to concentrate resources and efforts in a few areas only. The open nature of scientific inquiry makes it possible to learn rapidly from other scientists' results, thus offering the possibility of addressing other fields of science.

8.5. The Relationship between Size and Scientific Specialization

The degree of specialization across subfields of science is related in this section to the size of the national scientific efforts. This investigation follows the analysis developed in section 8.3 on the link between national degrees of technological specialization and the overall volume of their technological activities. The chi square values described above will be used as an indicator of the degree of scientific specialization of individual countries. As a proxy for the size of national scientific activity we will use the number of researchers and scientists employed in the non-business sector (i.e. higher education, plus government and non-profit institutions) in full-time equivalent units. This appears as the best proxy of the volume of resources devoted to science by each country, and to the number of authors of scientific papers, which is available from existing indicators. The average number

Degree of specialization and size of scientific activity -

Chi square values of the distribution of the number of papers by subfield of science and number of researchers and scientists, 1973-78, 1979-84

A1: Number of papers, 1973-78

A2: Number of papers, 1979-84

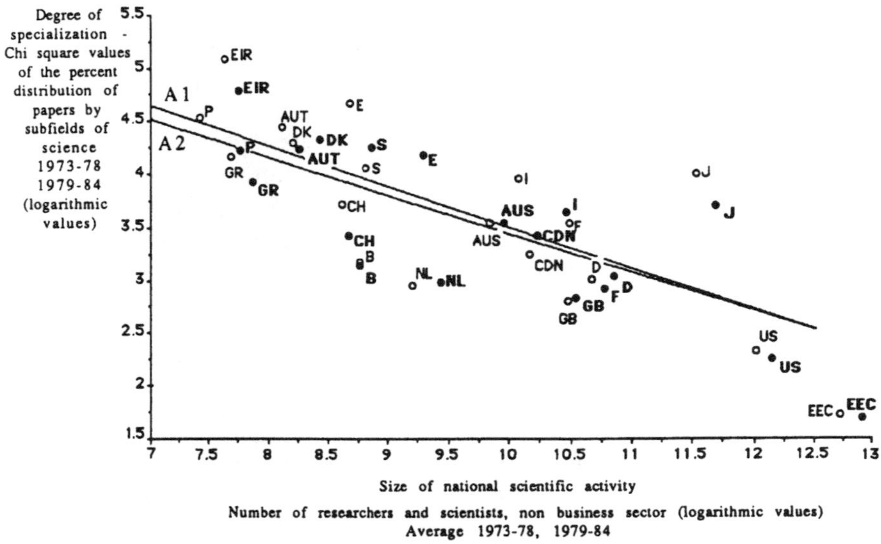

o normal = calculated on the number of papers, 1973-78

• bold = calculated on the number of papers, 1979-84

Legend

AUS= Australia	I= Italy
AUT= Austria	J= Japan
B= Belgium	NL= Netherlands
CDN= Canada	E= Spain
DK= Denmark	S= Sweden
F= France	CH= Switzerland
D= W. Germany	GB= Great Britain
GR= Greece	US= United States
EIR= Ireland	EEC= European Community

Fig. 8.4.

of researchers over the periods in which scientific papers are aggregated will be used.[9]

Figures 8.4, 8.5 and 8.6 plot the position of individual countries, showing the relationship between degree of specialization and size of the national scientific base. The two variables are transformed into their natural logarithm. Although a negative relation is evident in all graphs, the results obtained for bibliometric indicators are much less clear than for patents: the "size effect" seems to be less important for the degree of specialization in science than in technology. The regression line is drawn in order to illustrate the average pattern among all countries considered.

Figures 8.4 and 8.5 show that for both scientific papers and citations Japan and Italy, with few other countries, have a higher than expected degree of specializa-

Degree of specialization and size of scientific activity.

Chi square values of the distribution of the number of paper citations by subfields of science and number of researchers and scientists, 1973-78, 1979-84
A1: Number of paper citations, 1973-78
A2: Number of paper citations, 1979-84

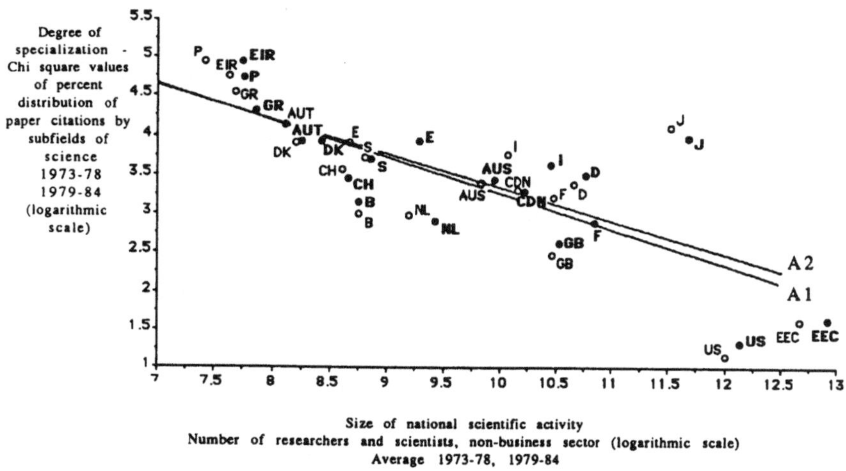

o normal = calculated on the number of paper citations, 1973-78
• bold = calculated on the number of paper citations, 1979-84

Legend

AUS= Australia	I= Italy
AUT= Austria	J= Japan
B= Belgium	NL= Netherlands
CDN= Canada	E= Spain
DK= Denmark	S= Sweden
F= France	CH= Switzerland
D= FR Germany	GB= United Kingdom
GR= Greece	US= United States
EIR= Ireland	EEC= European Community

Fig. 8.5.

tion, relative to the size of their scientific activity. Conversely, the US, the UK, the Netherlands, Belgium and Switzerland show the broadest distribution of their efforts across fields of science, relative to their size. The lowest degree of specialization is found for the EEC aggregate in the case of scientific papers and for the US in the case of citations. Over time, both distributions become slightly more uniform, suggesting a broader diversification of the areas of scientific research in most OECD countries.

For the second database, data for the two subperiods, 1981–83 and 1984–86 have been grouped together in order to show a more solid picture of the distribution of scientific activity in the 1980s. Figure 8.6 shows the degree of specialization measured by both papers and citations. The relative position of most countries is confirmed, with Japan showing by far the highest degree of specialization, followed by Italy and Spain, while the US, the UK, Canada, the Netherlands, Belgium and Switzerland show the broadest distribution of their scientific activity in relation to

Degree of specialization and size of scientific activity

Chi square values of the distribution of the number of papers and paper citations by subfield of science, 1981-86

BPAP: Number of papers, 1981-86

BCIT: Number of paper citations, 1981-86

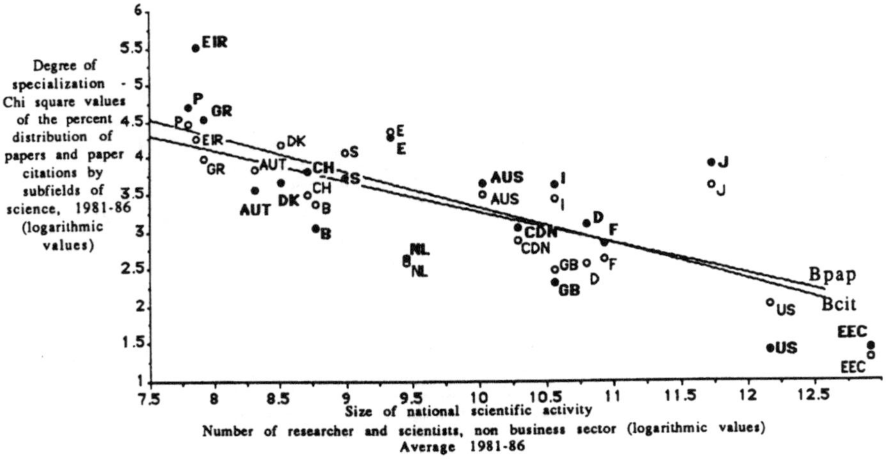

o normal = calculated on the number of papers, 1981-86

• bold= calculated on the number of paper citations, 1981-86

Legend

AUS= Australia	I= Italy
AUT= Austria	J= Japan
B= Belgium	NL= Netherlands
CDN= Canada	E= Spain
DK= Denmark	S= Sweden
F= France	CH= Switzerland
D= W. Germany	GB= Great Britain
GR= Greece	US= United States
EIR= Ireland	EEC= European Community

Fig. 8.6.

their size. The average distribution of papers is more uniform than that of paper citations, as pointed out above.

The results of this analysis show that, generally speaking, countries devoting the largest resources to science can afford to spread their efforts over a wider range of research areas. This relationship is, however, very loose, and even small countries show a high diversification of their scientific efforts. Japan and, to a lesser extent, Italy are the only countries showing consistently (over time, across the two databases, and for both paper counts and paper citations) a substantially higher degree of specialization in relation to the size of their scientific activity.

8.6. Diverging Trends in Science and Technology?

The evidence presented in this Chapter on the relationship between country size and degree of specialization in science and technology offers new insights into the processes already described in Chapters 2 and 3. First, we have already noted that a growing convergence of advanced countries emerges for several economic and technological variables, including productivity and S&T activities. Second, science and technology have undergone the same process of internationalization as is evident in economic activity. The findings on countries' specialization can now be related to these general processes.

A Summary of Technological Specialization. A few regularities can be identified from the data on technology:

i) countries devoting fewer resources to R&D tend to be more specialized in few technological sectors, and the degree of specialization is higher in terms of the impact of their technological activities than for the simple count of patent data;

ii) the degree of specialization increases over time and appears fairly stable when measured in different patent institutions and according to the various sectoral classifications.

Only large countries can afford to distribute their innovations more uniformly across technologies. Small countries, on the contrary, are to some extent forced to specialize in selected niches, suggesting that they are more dependent on international technology flows and cooperation than large ones. This result will come as no surprise to students of international economics, since the same relationship has long been shown for trade. Identifying these parallel patterns in trade and technology helps to define the context of the process of internationalization affecting all the advanced countries.

For some countries, the degree of specialization appears substantially higher than would be expected from their size. The most notable case is Japan, and, to a lesser extent, also Italy. Conversely, Great Britain and France have a comparatively low level of specialization. While the former countries have concentrated their efforts in few fields, for the latter the pattern of national technological accumulation has led to a more diversified range of activities. These differences can be seen as the outcome of diverging technological strategies followed by firms and governments, focusing on specialization and international integration in the former countries, and on national capabilities in the latter.

The increase over time in the degree of specialization in technology shown by most countries is parallel to the internationalization of technology markets stressed in Chapter 3. The drive to appropriate the returns from technological innovation in global markets has led to greater technology flows, with most countries developing a more specialized profile of S&T activities and accepting greater integration into the international technological system. This strategy is becoming more important for smaller countries, which are somehow forced to find their own "niches" not only in production and markets but also in technological activities. A necessary condition for the successful emergence of a "technological niche" is reaching some critical threshold of innovative resources in a specific sector. Together with a variety of other factors, ranging from technological accumulation to public policies, the definition of areas of clear technological specialization can contribute to improved national competitiveness in world markets. The evidence presented in this Chapter raises new questions on the possible link between the patterns of technological specialization and national performances. In Chapter 10 a further exploration of the impact of national specialization will be developed, relating it to the rate of growth of technological activities and industrial production in advanced countries.

The Results for Scientific Specialization. The relative convergence among countries in science indicators and the drop registered in the degree of specialization across sectors point to a process of growing similarity in the distribution of the activity of advanced countries across a few fields of science. This pattern is related to the internationalization of science, which facilitates the diffusion of and access to scientific knowledge in all fields, resulting also in an increasing cooperation among national scientific communities, as shown, for example, by the growing number of papers co-authored by scientists of different countries.

Countries appear to meet growing difficulties in expanding their scientific activity further in the fields where they are already strong. Such "diminishing returns" may lead to a greater diversification of national scientific efforts as the areas of previous weakness grow faster than the average. In fact the drop in specialization can hardly be considered the result of deliberate science policy in all countries (even if choices in this sense are made by some countries), as the cumulative nature of scientific knowledge and, more importantly, the institutional inertia of national scientific organizations would tend to reproduce and consolidate the traditional pattern of national activity. Such institutional aspects of the dynamics of science should be borne in mind as they strongly affect the patterns of scientific specialization.

Patterns in Science and Technology. The parallel results shown in this Chapter for scientific and technological indicators allow significant conclusions to be drawn. We are fully aware that the difference between "science" and "technology" is, in part, artificial, and that the two are increasinlgy integrated in modern innovative systems. However, the indicators considered, patenting and bibliometrics, do reflect two different aspects of knowledge and research. While patenting is related to activities of a proprietary nature, publications in journals are generally associated with the free transfer of knowledge and experience among different national scientific communities.

We have already pointed out the diverging trends in the degree of specialization for science and technology. Over time, most countries have increasingly focused their technological activities, measured by patents, in selected fields and spread their scientific efforts, measured by the number of papers, over a broader set of areas. Such opposing patterns can be explained by the different nature and dynamics of science and technology activities. A crucial element which tends to distinguish science from technology is the degree of openness of existing knowledge. Free access through scientific publications to the results and the advances of the scientific community is a necessary condition for the rapid growth of activity in the weaker countries and sectors. When a large part of the relevant knowledge is secret or proprietary, as in the case of technology, there are serious barriers to entry into new fields, and concentrating efforts in the areas of strength produces greater results than difersifying them.

The institutional differences between the science and technology systems play a major role here, and further comparative analyses of the changes in the patterns of specialization shown by science and technology indicators may be needed at a more disaggregated level in order to show more specific links between national efforts in science and technology.

Notes

1. By "degree of specialization" we mean an indicator of how evenly or unevenly the scientific or technological activities of a given country are distributed across all the sectors.
2. See, among a large literature, Kristensen and Levinsen (1983), Walsh (1988). This link has also been found for trade in high technology products. A recent study (Amendola and Perrucci, 1990) has found an inverse relationship between the degree of specialization of the distribution of national trade in 250 very detailed high technology product groups on the one hand, and the size of countries on the other.
3. In a work in progress report (Archibugi and Pianta, 1989), we have examined the standard deviations of the vectors of specialization indexes (TRCA) obtained for patents. However, the indexes of specialization have asymmetrical values (the value ranges from 0 to 1 in the case of a country's disadvantage, but from 1 to ∞ in the case of an advantage), and this is also reflected in the values of the standard deviation (see Grupp and Schwitalla (1989), Engelsmann and van Raan (1990)). The chi square values offer a better description of the countries' distribution of activities across sectors. The findings for both the indicators considered, however, draw a similar picture of the degree of specialization of the countries considered.
4. Since a country with a chi square value equal to 0 would not have any sectors of relative strength or weakness, its specialization indexes would be equal to 1 in all the classes.
5. In calculating the chi squares, the residual classes have been excluded; the values are calculated on 31 IPC classes (exluding the class Others) and on 41 SIC classes (exluding the classes Unclassified and Other industries).
6. Data on cumulative R&D expenditure are drawn from the OECD Main Science and Technology Indicators and are expressed in millions of US dollars at 1985 constant prices. For exchange rates, OECD purchasing power parities have been used. Missing values in the OECD series have been replaced by the estimated values of the regression line for the 1975–88 period. Similar results can also be obtained by using the R&D performed in the business sector as an indicator.
7. Portugal has been excluded from these graphs and the calculations of the regression lines due to the extremely high (and unreliable) value of its chi square.
8. While the original database includes 106 subfields, those with the smallest number of papers and citations have been merged into other classes, in order to assure statistically significant results in the analysis of sectoral specialization. The 8 fields of science are subdivided as follows: Clinical Medicine in 34 subfields; Biomedical research in 15 subfields; Biology in 10 subfields; Chemistry in 7 subfields; Physics in 9 subfields; Earth and Space sciences in 5 subfields; Engineering and Technology in 12 subfields; Mathematics in 4 subfields. The full statistical analysis of bibliometric data at this level of disaggregation is provided in Pianta and Simonetti, 1990.
9. OECD data from the Main Science and Technology Indicators have been used. The OECD series is, however, highly incomplete and some estimates had to be made. In evaluating this indicator, it should also be remembered also that the propensity of researchers to publish papers varies widely across fields and countries, for institutional and other reasons, and different behaviour can be found between university scientists and researchers in government and non-profit institutions. Furthermore some papers included in the data bases considered are authored by people not included in this group, especially in the business sector. For all these reasons, the data on researchers should be considered with caution. It should be remembered, however, that all the measures of the size of a country's scientific activity are strongly and positively correlated.

CHAPTER 9

The Inter-Industry Structure of Technological Specialization

9.1. The Degree of Specialization by Technological Sector

Three aspects of national technological specialization were shown in the previous Chapter: i) the existence of substantial differences among countries in the degreee of specialization, ii) the general increase over the 1980s, and iii) the national specificities, with the degrees of specialization that have grown faster in some countries, and decreased in a few others. We have not yet, however, taken into account the influence of individual technological sectors on the general degree of specialization. Obviously, significant differences exist across sectors and over time on the distribution of inventions by country. In some technological fields, the majority of countries have an innovative activity broadly proportional to their aggregate S&T size, while other fields can be deeply affected by the strengths (or weaknesses) of particular countries. Within each sector, the distribution across countries of technological activities may change over time, leading to a more even distribution of innovations, or to greater concentration in a few countries. These questions, which are strictly related to the analysis of the degree of specialization carried out in the previous Chapter, are addressed in this Chapter.[1]

The identification of sector-specific patterns of specialization has a direct relevance for policy analysis since deliberate strategies of both national governments and firms are often aimed at specific technological areas. If the production of innovations in a specific field is unevenly distributed across countries, it is likely that the achievement of a critical mass of resources might be a preliminary condition for entering certain areas, and policies for international integration might be effective. On the contrary, an even distribution of innovations across countries in a certain field would suggest that access to that technology is relatively easy, the nature of this knowledge is localized, and that policies for international cooperation are less likely to be relevant. Several elements might explain cross-technology differences in the process of specialization. The following factors are likely to be important in explaining the distribution of technological capabilities across countries.

i) The technological barriers to entry. Fields which require large investment and the mastering of complex knowledge systems are likely to show a less even distribution of technological activities across countries. When a substantial critical mass of resources is required for successful innovation, the distribution across countries is likely to correspond either to large government programmes or to the activities of large companies.

ii) The range of applications. While specialized technologies are generally concentrated in a few locations, generic or pervasive technologies are needed by all the national innovation systems.

iii) Finally, the existence of national differences in the structure of final demand may influence the international distribution of technological capabilities.

Since pervasiveness might be a key issue in explaining specialization across technologies, it should be borne in mind that it can be referred alternatively to producers or users (see Archibugi, 1988b). A technology such as computers, for example, is highly pervasive in use but strongly concentrated in production in a few oligopolistic companies based in selected countries only. The indicator employed here accounts for the production of technology only, and provides no information about users.

9.2. Method of Analysis

The aim of the analysis of the technological specialization in individual sectors is to identify the existing patterns, their relative shifts over time and to outline a tentative typology. The indicator used in this Chapter is again patents granted and patent citations in the US, according to the 2-digit SIC classes already employed in Chapter 5. We will not use other patent indicators (drawn from other patenting institutions, such as the EPO, and based on the IPC classification) since we want to identify the shifts over time which can be investigated on the time series available for US patents only.

As in Chapter 8, we are again considering a matrix where the rows show the SIC classes and the columns the individual countries, and where each cell shows the number of patents granted (or of citations received) by country and technological sector. The degree of specialization of each technology across countries will be measured here, as in the previous Chapter, by using the chi square values. While in Chapter 8 the degree of specialization was investigated for each country by comparing the columns (which indicate national percentage distributions) to the total distribution for the world, here the approach is reversed, and the rows of the percentage distribution by sector across countries is compared to the percent distribution by sector of the total distribution for the world. If all the countries have a share of patents (or citations) in a given class equal to their total share of patents (i.e. in absence of specialization), the chi square value of this class is equal to zero.

9.3. The Degree of Specialization of Individual Technologies

Table 9.1 reports the chi square values for each class over the two periods for both patents and patent citations (the classes of Ordnance and Guided missiles have here been merged because of the small number of patents granted). The value of the chi square for the overall matrix increases between 1975–81 and 1982–88 from 539 to 704 for patents and from 582 to 878 for patent citations. Patent citations

TABLE 9.1.

TECHNOLOGICAL SPECIALIZATION BY SECTOR ACROSS COUNTRIES

Chi square values by product class for patents granted and patent citations in the US
a) Patents granted in the US, 1975-81 and 1982-88
b) Patent citations in the US, 1975-81 and 1982-88

SIC classes	a) Patents granted Chi square 1975-81	1982-88	Rates of change %	b) Patent citations Chi square 1975-81	1982-88	Rates of change %
1 Food.Kindred Products	6.2	13.12	111.64	6.44	21.45	232.83
2 Textile Mill Products	5.38	4.56	-15.24	3.51	3.14	-10.59
3 Inorganic Chemicals	2.43	4.98	103.13	1.31	8.28	532.73
4 Organic Chemicals	9.6	8.01	-16.57	10.29	13.38	29.98
5 Plastic Matrls,Synth Res	7.45	3.85	-48.28	10.19	3.49	-65.71
6 Agricultural & other Chem.	13.96	17.25	23.61	17.04	31.01	81.94
7 Soaps,Detergents,Clnrs	8.15	24.22	197.13	18.78	67.00	256.83
8 Paints,Allied Chemicals	2.43	2.73	12.45	2.61	3.00	14.84
9 Misc Chemical Products	3.97	3.21	-19.28	4.18	4.57	9.46
10 Drugs & Medicines	9.44	8.54	-9.52	10.33	14.37	39.14
11 Petrol,Nat Gas Extr,Ref	24.08	29.51	22.56	23.61	34.23	44.95
12 Rubber,Misc Plast Prods	0.62	0.86	38.93	0.68	0.92	35.05
13 Stone,Clay,Glass,Concr	1.99	0.84	-58.06	1.23	1.32	7.75
14 Primary Ferrous Prods	13.87	9.08	-34.52	12.05	6.13	-49.1
15 Prim,Sec Non-Ferr Prods	8.81	5.91	-32.86	6.91	8.68	25.63
16 Fabricated Metal Prods	5.1	6.08	19.2	5.27	5.89	11.78
17 Engines & Turbines	5.2	9.52	83.08	8.8	24.57	179.22
18 Farm,Garden Mach & Equip	12.45	21.71	74.26	13.87	23.63	70.39
19 Cnstr.Mng.Metal Hand Eqp	6.58	10.75	63.47	6.62	9.44	42.69
20 Metal Working Mach,Equip	2.77	5.14	85.95	3.00	5.54	84.67
21 Office Comput,Acctg Mach	3.86	18.1	368.32	4.16	10.17	144.34
22 Spec Ind Mach(exc M Wrk)	6.78	13.66	101.55	10.34	21.26	105.60
23 Genrl Indust Mach,Equip	1.45	2.1	44.88	1.64	3.28	100.15
24 Refrig,Servc Indust Mach	2.92	5.21	78.49	3.79	8.37	121.05
25 Misc Mach (exc Electric)	3.75	2.93	-21.76	18.46	19.32	4.65
26 Electr Trans,Distr Equip	1.47	0.97	-33.9	1.24	2.34	88.97
27 Electr Indust Apparatus	2.89	4.22	45.84	2.72	3.74	37.76
28 Household Appliances	5.2	5.01	-3.68	6.77	4.83	-28.63
29 Electr Lightng,Wirng Eqp	5.46	6.03	10.35	4.91	10.35	110.72
30 Misc Elec Mach,Eqp,Suppl	4.7	2.51	-46.59	6.56	3.59	-45.22
31 Radio.TV Receiving Equip	32.84	28.67	-12.7	22.47	19.32	-14.00
32 Elect Cmp,Acc,Comm Equip	4.71	5.43	15.41	4.23	3.45	-18.48
33 Motor Veh,Motor Veh Eqp	2.93	13.34	355.65	8.62	31.37	263.84
34 Ordnance-Guid Mssls *	15.00	24.28	61.91	17.34	32.47	87.23
35 Ship.Boat Bldng & Repair	10.51	24.13	129.6	15.86	20.79	31.12
36 Railroad Equipment	3.74	14.41	285.35	3.19	18.91	492.16
37 Motorcycles,Bicy & Parts	9.44	18.21	92.95	17.3	40.2	132.42
38 Misc Transportation Eqp	7.12	4.74	-33.5	3.6	8.42	133.91
39 Aircraft & other Mech, Parts	6.54	9.95	52.04	14.55	29.99	106.12
40 Prof,Scien Instruments	2.82	1.87	-33.74	1.68	1.35	-19.94

Sources: CNR-ISRDS elaboration on CHI Research data
* The classes of Ordnance and Guided Missiles have been merged

show a generally higher degree of specialization than patents: relevant innovations produced in each sector (as measured by patent citations) are more unevenly distributed across countries than the number of patents registered; a finding which complements the results obtained in the previous Chapter at the country level.

Increasing specialization is found in 25 (out of 40) classes for patents, and in 32 classes for citations. In 8 classes a reduction of the specialization for patents is married to an increase of the specialization for citations, confirming (as already stressed in Chapter 8) that a more uniform distribution of patents does not imply a more uniform distribution of their impact. In a few classes specialization decreases according to both patents and citations.

The degree of specialization in each sector is directly related to the activities of large firms as well as to national government policies. In commenting upon this results, we will refer to a break-down of patenting by sector and by government agencies, large and small firms provided, on the basis of a different classification, by Patel and Pavitt (1989a). In some cases, an even distribution of innovations across countries is to be found in sectors of moderate activities of large firms. This seems to be the case of metal products, textiles and other traditional sectors. Converserly, the uneven distribution of innovations across countries in consumers' electronics is associated to the high concentration of patenting in large firms. A high concentration of innovations in large firms, however, does not necessarily lead to an internationally uneven distribution of activities. This is the case of some chemicals, characterized by a plurality of large firms, which show a rather uniform geographical distribution and a more even distribution of patenting activities in the field. In other cases, the role of government agencies is the main determinant of uneven distribution, as in the case of Ordnance and missiles and other state-supported high-tech research programmes.

In order to examine whether sectors show a rising or falling degree of specialization, the rates of change between the chi squares obtained in the first and the second periods have been calculated and are presented in Table 9.1. A graphic comparison of the chi square values obtained for the two periods in all the sectors is offered in Fig. 9.1. A regression line is drawn, and its positive inclination indicates that a stable pattern of specialization was experienced in the majority of the technological classes: sectors with high degrees of specialization in the first period tend to show high chi square values also in the second and viceversa. This stable international distribution of activities lends additional support to the hypothesis of technological accumulation. On the other hand, the regression line lies above the 45 degrees line, showing the general trend towards increasing specialization in the overall distribution.

9.4. Sectoral Results

The specialization process is specific to each technological class. We will consider sector-specific aspects such as the degree of specialization, the main tendencies over time, and the contributions from individual countries. Figure 9.2 plots the chi square values obtained for patents in the 1982–88 period versus their rates of change from 1975–81 to 1982–88. Overall, no sector with high chi square values shows a drop in its degree of specialization (the single exception is represented by Radio & TV equipment, which still has a very high value). This result indicates that sectors with a high degree of specialization tend to increase it. On the other hand, sectors of low specialization are more likely to decrease it further; the diffusion of generic technologies leads to a more even international distribution of their production.

The degree of specialization in technological sectors
Chi square values, 1975-81 versus 1982-88,
SIC classes, patents granted in the US

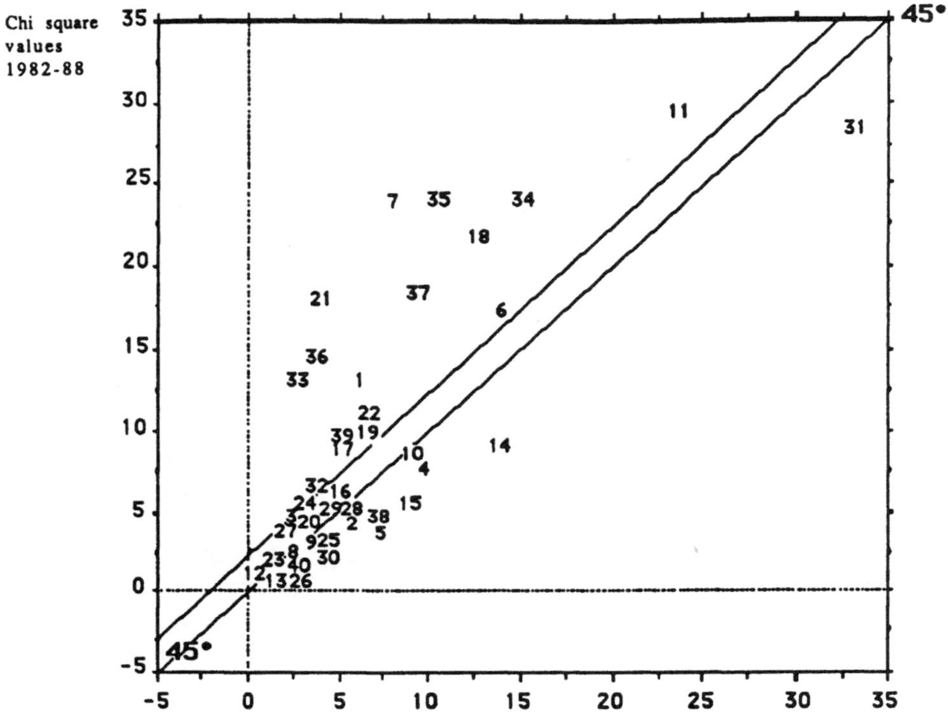

Legend

1 Food, Kindred Products
2 Textile Mill Products
3 Inorganic Chemicals
4 Organic Chemicals
5 Plastic Matrll, Synth Res
6 Agric. & Oth. Chemicals
7 Soaps, Detergents, Cnrs
8 Paints, Allied Chemicals
9 Misc Chemical Products
10 Drugs & Medicines
11 Petrol, Nat Gas Extr, Ref
12 Rubber, Misc Plast Prods
13 Stone, Clay, Glass, Concr
14 Primary Ferrous Prods
15 Prim, Sec Non-Ferr Prods
16 Fabricated Metal Prods
17 Engines & Turbines
18 Farm, Garden Mach & Equip
19 Cnstr, Mng, Metal Hand Eqp
20 Metal Working Mach, Equip
21 Office Comput,Acctg Mach
22 Spec Ind Mach(exc M Wrk)
23 Genrl Indust Mach,Equip
24 Refrig,Servc Indust Mach
25 Misc Mach (exc Electric)
26 Electr Trans,Distr Equip
27 Electr Indust Apparatus
28 Household Appliances
29 Electr Lightng,Wirng Eqp
30 Misc Elec Mach,Eqp,Suppl
31 Radio,TV Receiving Equip
32 Elect Cmp,Acc,Comm Equip
33 Motor Veh,Motor Veh Eqp
34 Guid Mssls,Spce Veh,Prts
35 Ship,Boat Bldng & Repair
36 Railroad Equipment
37 Motorcycles,Bicy & Parts
38 Misc Transportation Eqp
39 Aircraft & other Mech. Parts
40 Prof,Scien Instruments

Fig. 9.1.

Degree of specialization by technological sectors versus rates of change 1975-81 to 1982-88

Chi square values by SIC classes - Patents granted in the US

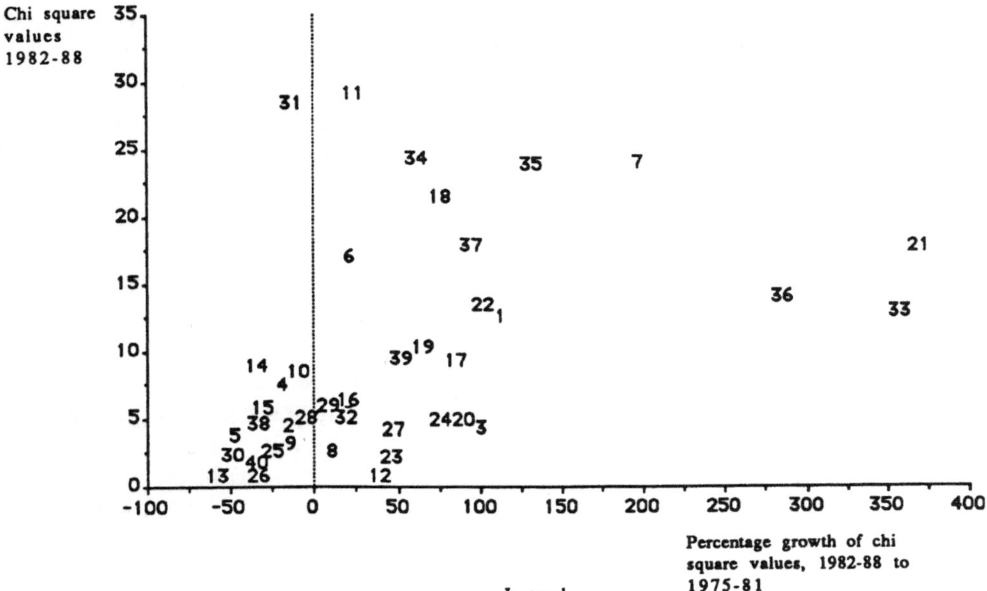

Legend

1 Food, Kindred Products
2 Textile Mill Products
3 Inorganic Chemicals
4 Organic Chemicals
5 Plastic Matrll, Synth Res
6 Agric. & Oth. Chemicals
7 Soaps, Detergents, Cnrs
8 Paints, Allied Chemicals
9 Misc Chemical Products
10 Drugs & Medicines
11 Petrol, Nat Gas Extr, Ref
12 Rubber, Misc Plast Prods
13 Stone, Clay, Glass, Concr
14 Primary Ferrous Prods
15 Prim, Sec Non-Ferr Prods
16 Fabricated Metal Prods
17 Engines & Turbines
18 Farm, Garden Mach & Equip
19 Cnstr, Mng, Metal Hand Eqp
20 Metal Working Mach, Equip
21 Office Comput,Acctg Mach
22 Spec Ind Mach(exc M Wrk)
23 Genrl Indust Mach,Equip
24 Refrig,Servc Indust Mach
25 Misc Mach (exc Electric)
26 Electr Trans,Distr Equip
27 Electr Indust Apparatus
28 Household Appliances
29 Electr Lightng,Wirng Eqp
30 Misc Elec Mach,Eqp,Suppl
31 Radio,TV Receiving Equip
32 Elect Cmp,Acc,Comm Equip
33 Motor Veh,Motor Veh Eqp
34 Guid Mssls,Spce Veh,Prts
35 Ship,Boat Bldng & Repair
36 Railroad Equipment
37 Motorcycles,Bicy & Parts
38 Misc Transportation Eqp
39 Aircraft & other Mech. Parts
40 Prof,Scien Instruments

Fig. 9.2.

Electrical-Electronic Sectors. A very large and pervasive sector, Professional and scientific instruments, has one of the lowest and decreasing specializations, a pattern shared by several classes containing a high number of patents. The majority of the electrical and electronic classes show a substantially stable degree of specialization, as indicated by the moderate values of the rates of change of the chi squares. The exceptions are the classes of Electrical transmission and Electrical industrial apparatus. However, these two classes have a low chi square value and are therefore sensitive to small variations. In Electrical lighting there is a clear tendency towards a decrease in the specialization for citations, principally due to the Japanese despecialization in the sector. Office computing has a pattern differing substantially from the other electronic classes: the chi square value is high, and has increased faster than in any other class (due especially to the very high number of patents granted to Japan). The impact of the inventions in this technology, however, seems to be more uniformly distributed across countries, as indicated by the lower value of the chi square for patent citations.

Chemicals and Allied Classes. Among the chemical classes two pervasive technologies, i.e. Paints & allied and Plastic materials, show very low degrees of specialization. The same applies to sectors such as Drugs, Organic chemicals and Inorganic chemicals, which are characterized by a high propensity to patent. An opposite pattern of growing specialization is shown by Agricultural and other chemicals and Soap & detergents, a result which is affected by the worsening of Japanese activity. The high degree of specialization of the latter class (which ranks first according to the degree of specialization in citations) is also due to the high number of patents granted to Great Britain and Belgium. In Food and kindred products the high specialization is connected to Japanese and German weakness, and to Danish strength. Classes connected with mineral products are found among those with low specialization, especially the class of Stone, clay, glass. The only exception is represented by Petroleum and natural gas, which is strongly influenced by the above average share held by the United States (in its domestic market) and by the low activity of Japan and Germany.

Machinery. The machinery classes constitute the bulk of the technological classes with medium-high and growing specialization. This result confirms the differentiated and localized expertise embodied in machinery, which is transferred to other sectors according to specific user-producers interrelations described by several case studies (Lundvall (1985), von Hippel (1988)). Japanese strength in Engines and turbines, and its weaknesses in Farm machinery, Construction and mining machinery influence the results for the overall distribution. The uneven distribution of Metal working machinery and Specialized industrial machinery is due, in particular, to German and Swiss strength.

Transport. The impressive increase in specialization in Motor vehicles and in Motorcycles has occurred as a consequence of the increased number of Japanese patents. In Railroad equipment and Shipbuilding, on the contrary, the increase in specialization is due to the prominent activity of several countries: Germany, the United Kingdom and Sweden for the former, Canada and Australia for the latter.

Protected High-Technology Sectors. The degrees of specialization of two high technology and protected sectors, i.e. Aircraft and Ordnance & missiles, are basically related to the low number of Japanese patents. In both classes the values obtained with patent citations are considerably higher than those for patents.

Miscellaneous Classes. The four residual classes, i.e. Miscellaneous chemical products, Miscellaneous electrical machinery, Miscellaneous non-electric machinery and Miscellaneous transportation present low and decreasing degrees of specialization. Since they contain heterogeneous inventions, it is likely that the dynamics of specialization is somewhat concealed by the classification employed. As already suggested in the previous Chapters, specialization across both technologies or countries seems to increase when a specific and disaggregated classification is considered.

9.5. Discussion

This Chapter has addressed the question of the increasing degree of specialization from a perspective mirroring that presented in Chapter 8, i.e. focusing on technological classes instead of countries. Since we have employed the same matrixes of patents and patent citations by country and SIC classes, we have obviously found the same general tendency towards increasing specialization. However, the generalized increase of specialization is shown to have affected single technological fields in a differentiated manner.

While the evidence presented above does not allow a clear-cut technological typology to be made, a few conclusions can be drawn. Most technologies with high and increasing specialization are associated with those sectors which might be labelled "production intensive", following Pavitt's taxonomy (Pavitt, 1984). The bulk of these classes includes specialized machinery and equipment, the kind of technologies which are highly specific to a country's industrial structure and to individual sectors. The same high specialization can be found in "scale intensive" technologies, such as in transport and chemical processes, with the same pattern shown also by computers. Production in these areas tends to be characterized by a high degree of industrial concentration, coupled with an uneven distribution of technological activities across countries.

The most significant results, however, have emerged for the classes which do not follow the general pattern. The fields with a low level of specialization generally show a further decrease over time. It should be noted, however, that all the miscellaneous classes have both low and decreasing degrees of specialization. The classification adopted is clearly responsible for this result. In other words, it is likely that if a more disaggregated classification had been adopted, the shift towards increasing specialization, which is one of the core issues of this study, would have been more significant.

The same group with low and/or declining specialization includes traditional and generic technologies, such as Textiles, Household appliances, and Metal products. In these fields technological entry barriers are low, and we would expect patenting activity to spread more easily across countries. However, the classes with low and declining specialization also include very pervasive sectors, as in the case of Professional and scientific instruments and Plastic materials. The classes belonging to the electronic cluster, with the notable exception of Office computing, are to be found among the sectors with stagnant/declining and low specialization. This result seems to confirm the pervasive nature of electronic technologies. Their application to a variety of fields may have further increased their even distribution across countries.

In conclusion, the analysis of the degree of specialization at the sectoral level seems to bear out the results already presented at the country level inasmuch as it indicates that the decrease of specialization is confined to miscellaneous, generic and very pervasive sectors. On the contrary, the increases in the degree of specialization are associated with fields where the international division of labour among countries is also growing. This pattern in turn provides a broad picture which makes it possible to identify the fields where policies for international integration and cooperation are likely to be more relevant and effective.

Note

1. This Chapter draws upon work carried out within our research group by Roberto Simonetti (1992).

CHAPTER 10

The Patterns and Impact of Technological Specialization

10.1. The Nature and Effect of Technological Specialization

The findings of the previous Chapters have described the process of specialization in science and technology in advanced countries. Quantitative data on several S&T indicators and databases have been used to examine the trends of the 1970s and 1980s at national and at sectoral level. This Chapter will pull together and summarize the different threads of the analysis and go on to examine the relationship between specialization and national performance. After the analysis in the previous Chapters of the specialization profiles of advanced countries in technology and science and the data provided on the degree of specialization in countries and sectors, the concluding questions to be addressed focus on the different national patterns of technological specialization and on the impact they have on national performance. In particular, two issues will be examined:

i) The first question is: how can the similarities and differences in national patterns of technological activity be defined? Are there common patterns across countries emerging from the comparison of national sectoral specialization?

One of the main results of the literature on national innovation systems (reviewed above in Chapter 2) is that countries differ in the methods used to produce, introduce and disseminate innovations in their economic and social systems. These differences also emerge in their distribution of activities across fields, as shown in Chapters 4 to 7. The analysis developed in the following section brings together the different patterns shown by individual countries, so as to present a picture of the similarities and differences of pairs of countries across the spectrum of technologies. This serves to highlight the different "paths of specialization" and the strategies followed by large and small countries, resulting in what may be described as the international "division of labour" in technology. These data will show whether some countries are "converging" to a similar pattern of specialization, resulting in greater international competition, or whether they are specializing in different fields, leading to greater integration and interdependence in world technological activities.

ii) The second question is: what is the impact of national patterns of technological specialization on technology and economic indicators? What regularities can be found between particular patterns of national specialization and the characteristics of national technological efforts and economic performance?

In section 10.3 this second investigation explores the relationship between the different characteristics of national technological activities (including the countries' size, nature and degree of specialization, etc.) and relates them to the economic and technological performance of advanced countries. This provides an examination of the effect specific patterns of technological specialization have on national performance.

10.2. Similarities and Differences in National Technological Specialization

The description of national patterns of specialization in technology in Chapters 4 to 6 has already pointed out the different characteristics of the sectoral activity of advanced countries. However, a more formalized investigation of how similar or different countries are in their technological specialization may provide a better picture of the extent to which national activities overlap, converging on a similar sectoral profile, or focus on different fields, leading to growing integration. The aim of this analysis is to explore whether different "models of specialization', or "paths of technological development" can be identified for separate groups of countries. The analysis of how these similarities and differences have evolved over time also provides new information on the dynamics of the specialization process.

Much conventional wisdom suggests that as countries develop, they tend to reach a similar pattern of technological activity, with greater resources devoted to innovative activites, entering new high technology fields, and assuming a specialization profile typical of "advanced" countries, which is largely modeled on the US. In addition to the convergence in terms of aggregate resources devoted to innovative activities, which is documented in Chapter 3, such conventional views suggest that a similar process of convergence is also to be expected for the sectoral distribution of technological efforts. The closing of the "technology gap" would imply a move towards a standard composition of national innovative activities where "high tech" fields are largely represented.

An alternative view has been proposed by the literature on national innovation systems. It has been pointed out that there are significant differences across countries in the way they introduce and diffuse innovations. In this book we have further developed this issue, suggesting that the closing of the technology gap has lead to an increasing specialization and diversity of advanced countries. The qualitative differences in national systems are reflected in the quantitative results of their innovative activities. The profiles of national technological specialization reported in Chapters 4 to 6 show, in fact, that each country has well defined strengths and weaknesses in different areas. More importantly, in Chapter 8 it was shown that almost all countries have experienced growing sectoral specialization, which suggests an inceasing specificity of national profiles with respect to the world pattern of technological activity. Such findings may also lead us to expect substantially different national patterns for countries with different sizes and degrees of technological intensity.

In this section we further explore this question by examining the similarities and differences between pairs of countries. Of particular interest here is the comparison of the patterns of specialization of the EEC, the US and Japan, showing whether the three major regions of the advanced world are becoming more or less similar in terms of sectoral pattern of technological activities. The two possible patterns one may expect to find in this analysis are as follows: on the one hand, a convergence of several countries, including small and medium-sized ones, upon a distribution of their activities covering a large number of fields, with growing similarities to large advanced countries; on the other, more specific national specialization profiles, with a concentration of technological activities in a few sectors only and greater reliance on integration in the world technology market in other fields.

These two patterns may be associated with different policy approaches. The former may be the result of policies directing national efforts into several advanced technological sectors. In this case, countries choose to compete in several areas, relying on a variety of macroeconomic and industrial policy tools supporting the existing technological activities. The latter may be related to policies emphasizing efforts in the fields of greater national strength, and access to foreign know how in the areas of weakness, assuming growing flows of technology across borders. In this case, the search for international competitiveness targets carefully selected "niches" where national resources are concentrated in order to achieve international excellence.

Obviously, a high level of national technological activity is an important condition for the emergence of the former pattern, while smaller countries may be forced to follow the latter. Since large countries produce innovations in the majority of fields, they are more likely to compete over a large range of sectors. In order to address these questions empirically, the countries' sectoral distribution of technological activities has been compared, developing a measure of distance between pairs of countries. In Chapter 8 a country's sectoral distribution of science and technology activities was compared to the world profile in order to identify the degree of national specialization. The same approach can also be used to compare pairs of countries, thus describing similarities and differences between individual countries. In this section the same chi square method used above to measure the degree of specializion (relatively to the world pattern) is employed to measure the distance between pairs of countries. This analysis is developed for technological specialization only, leaving aside the dynamics of scientific activity, which is characterized by different patterns, as seen in Chapter 8. For technological activities, much detailed information has already been provided on the countries' sectoral distribution of patents (in Chapters 4, 5 and 6) and the evidence of the previous sections has indicated a well defined dynamics of growing sectoral specialization.

The distance between individual countries will be calculated for the periods 1975–81 and 1982–88, using data on the sectoral distribution across 41 SIC classes of patents granted in the US.[1] The method used is the following. The percentage distribution of patents in 41 non-residual SIC classes (see Table 5.3 for the listing) is considered for two countries, thus making the measure independent of the number of a country's patents. The square of the difference between the percentages for each class is calculated and divided by the share the class holds of total world patents. The sum of these weighted squared differences is then divided by the maximum possible value of the the distance between two countries, resulting in an indicator varying between 0 and 1,000, which standardizes the values and allows comparisons over time. The formula of the distance index is the following:

$$D_{ab} = [\sum_{i=1}^{n}[(p_{ia} - p_{ib})^2/p_{iw}]]/D_{\max} * 1000$$

where:

D_{ab} is the distance index between country a and country b, which varies between 0 and 1000;

p_{ia} is the percentage of patents of country a in sector i;

p_{ib} is the percentage of patents of country b in sector i;

p_{iw} is the percentage of patents of the world total in sector i;

n is equal to 41 non-residual SIC classes.

$D_{\max}$ is the maximum value of the distance for a given world distribution in n classes.[2]

The distance index between two countries is equal to zero when they have the same percentage distribution of patents across classes, and it grows rapidly when one country is strong in fields where the other holds few or no patents. The standardized index is a modification of the simpler analysis of specialization developed in Chapter 8, and is needed to provide a benchmark for the comparison of different pairs of countries and to assess the distances between two countries relatively to varying sectoral distributions of world patents. Table 10.1 shows the distance between the three major regions, the EEC, the US and Japan, for the late 1970s and the mid–1980s.

The EEC and the US show a closer sectoral distribution of patenting activities and become slightly more similar over the two periods (distance indexes of 11.2 and 9.5). In 1975–81 Japan shows a greater difference from both the US (index equal to 22) and the EEC (index equal to 24); in the second period Japan increases its difference from both areas substantially, and in particular from the EEC (index equal to 30). These figures show to what extent Japan is a country with a very distinct specialization in technology, with a growing distance from the US and Europe. The EEC and the US, on the other hand, are rather similar and in the 1980s they even increased their similarity (the parallels existing in their S&T institutional system may have played a role in this). The data for the EEC (resulting from the aggregation of the twelve countries) may also suggest a broader picture of technological activities, as the different national specializations are leveled off in the distribution.

Table 10.2 provides the same distance indexes for 13 countries; the smallest patenting countries (Ireland, Portugal and Greece) are excluded because the low number of patents they hold makes the analysis unreliable. The pattern of relative differences emerging from the matrix is fairly complex. The US most resembles France, the UK and Canada in both periods. In turn, France and the UK are very close and are similar to the US and Germany. The sectoral distribution of patents by Germany is closer to France, the UK and Italy. The major similarities for Italy are first with Germany, and then with France, the UK and Switzerland, showing

TABLE 10.1.

DISTANCE AMONG THE PATTERNS OF TECHNOLOGICAL SPECIALIZATION OF ADVANCED COUNTRIES
The distance indicator is based on the percent distribution across 41 SIC classes of patents granted in the US to individual countries. The index varies from 0 to 1000.
Distance values are divided by the maximum possible distance, calculated on the average of the five more extreme cases.
Period 1: 1975-81
Period 2: 1982-88

Countries	Period	USA	Japan	EEC
USA	1	-	22.33	11.18
	2	-	24.64	9.52
Japan	1	22.33	-	24.39
	2	24.64	-	30.22
EEC	1	11.18	24.39	-
	2	9.52	30.22	-

Source: CNR-ISRDS elaboration on CHI Research data

higher distance indexes.

Japan has generally high values of the distance index, and appears to be closer to the US and France. The remaining countries show very high indexes and tend to show closer similarities either to the US or to the major European countries.

In interpreting such patterns of relative differences across countries, the impact of countries' size of patenting activities is evident. The larger countries, and in particular the US, France, the UK and Germany, appear rather close, as they distribute their patents across all sectors. A similar broad distribution is shown by Japan, which, in spite of its greater distance, proves to be closest to some countries of the above group.

Strong differences emerge on the other hand for small countries, which concentrate their patents in few (and different) areas. Smaller countries differ greatly from one another and tend to be closer to the larger country with which they share a particular specialization. In other words, smaller countries appear to have developed a technological specialization in selected, country-specific "niches", which are, however, very different across countries. This finding is consistent with research on the determinants of innovation in small countries (see Walsh, 1988; Kristensen and Levinsen, 1983).

The marked diversity of the specialization patterns of small countries suggests that a limited national S&T budget does not prevent the development of activities in a broad variety of technological sectors, the major exception being those state-supported and military-oriented fields where large government expenditure is needed to sustain the development of such capabilities. A small country size means

TABLE 10.2

DISTANCE AMONG THE PATTERNS OF TECHNOLOGICAL SPECIALIZATION OF ADVANCED COUNTRIES
The distance indicator is based on the percent distribution across 41 SIC classes of patents granted in the US to individual countries. The index varies from 0 to 1000.
Distance values are divided by the maximum possible distance, calculated on the average of the five more extreme cases.
Period 1: 1975-81
Period 2: 1982-88

Countries	Period	USA	Japan	W. Germ.	France	U.Kingdo	Italy	Netherl.	Belgium	Denmark	Spain	Switzerl	Sweden	Canada
USA	1	-	22.33	17.40	8.95	12.25	37.00	28.82	46.71	36.13	68.51	56.40	31.85	14.19
	2	-	24.64	17.11	6.89	11.02	35.18	23.36	45.11	37.20	69.06	46.09	29.75	13.84
Japan	1	22.33	-	29.28	23.70	25.79	50.74	43.18	57.48	56.14	104.10	73.21	68.58	56.14
	2	24.64	-	37.93	25.61	30.18	60.86	35.26	85.40	79.02	104.38	76.13	64.80	79.02
W. Germany	1	17.40	29.28	-	9.62	9.51	13.36	54.04	31.20	42.27	45.05	27.89	43.58	31.89
	2	17.11	37.93	-	12.24	14.74	13.58	48.81	46.09	132.19	40.03	24.02	29.18	28.87
France	1	8.95	23.70	9.62	-	3.24	23.22	32.78	40.09	39.13	51.03	40.04	39.63	22.76
	2	6.89	25.61	12.24	-	5.09	26.53	25.47	41.04	40.85	55.71	35.78	33.78	21.85
U.Kingdom	1	12.25	25.79	9.51	3.24	-	26.44	78.24	37.50	33.77	53.78	39.02	42.40	27.86
	2	11.02	30.18	14.74	5.09	-	24.37	37.15	33.43	35.33	56.52	32.14	38.55	28.65
Italy	1	37.00	50.74	13.36	23.22	26.44	-	65.36	42.30	44.93	40.49	29.08	63.98	45.67
	2	35.18	60.86	13.58	26.53	24.37	-	64.82	53.61	41.20	38.19	19.79	44.79	42.18
Netherlands	1	28.82	43.18	54.04	32.78	78.24	65.36	-	85.78	75.04	105.82	82.71	79.91	46.56
	2	23.36	35.26	48.81	25.47	37.15	64.82	-	65.38	74.14	119.77	74.82	73.68	41.17
Belgium	1	46.71	57.48	31.20	40.09	37.50	42.30	85.78	-	64.95	84.88	47.47	86.04	68.43
	2	45.11	85.40	46.09	41.04	33.43	53.61	65.38	-	59.65	90.47	53.45	77.72	61.77
Denmark	1	36.13	56.14	42.27	39.13	33.77	44.93	75.04	64.95	-	57.73	53.34	41.58	41.24
	2	37.20	79.02	132.19	40.85	35.33	41.20	74.14	59.65	-	48.95	47.76	36.32	36.09
Spain	1	68.51	104.10	45.05	51.03	53.78	40.49	105.82	84.88	57.73	-	51.99	63.30	55.79
	2	69.06	104.38	40.03	55.71	56.52	38.19	119.77	90.47	48.95	-	60.10	42.30	56.86
Switzerland	1	56.40	73.21	27.89	40.04	39.02	29.08	82.71	47.47	53.34	51.99	-	76.81	79.79
	2	46.09	76.13	24.02	35.78	32.14	19.79	74.82	53.45	47.76	60.10	-	87.65	64.02
Sweden	1	31.85	68.58	43.58	39.63	42.40	63.98	79.91	86.04	41.58	63.30	76.81	-	20.94
	2	29.75	64.80	29.18	33.78	38.55	44.79	73.68	77.72	36.32	42.30	87.65	-	17.11
Canada	1	14.19	56.14	31.89	22.76	27.86	45.67	46.56	68.43	41.24	55.79	79.79	20.94	-
	2	13.84	79.02	28.87	21.85	28.65	42.18	41.17	61.77	36.09	56.86	64.02	17.11	-

Source: CNR-ISRDS elaboration on CHI Research data

that there is a need to specialize somewhere, but there is no general constraint on the particular fields where a country will have to concentrate.

The impact of common national specialization in selected sectors is evident from the distance matrix of Table 10.2. The strong similarity between France, the UK and, to a lesser extent, the US is due to the common importance of military-related areas (such as Aircraft, Guided missiles, Ordnance) and of classes such as Drugs and medicines and Agricultural and other chemicals. Some of these fields are areas of technological specialization also for Germany, which on the other hand shares with Italy and Switzerland a relative concentration of its patenting in Specialized industrial machinery and in some chemical classes. Other links can be identified among national specializations. A relative importance of electronic-related fields is at the root of the similarities shown by Japan, the US, France and the Netherlands. The US is also fairly close to Canada, which in turn shows a specialization profile similar to that of Sweden, with Denmark not too far removed. In these linkages the relative importance of natural resources and agriculture-related activities, as well as specialized production in fields such as shipbuilding and engineering, appears to be crucial.

The analysis of the changes over time in the distance indicators is of particular interest as it highlights the different "paths of specialization" shown by groups of countries. From the late 1970s to the mid–1980s, the US, the UK and France further increased their already strong similarity. Germany, on the other hand, moved away from this group, even though it remains close to it. Germany continues to be highly similar to Italy and is now closer to Switzerland and Sweden. In turn, Italy moved closer to the technological specialization of Switzerland and Sweden, while increasing its distance from most other countries.

Japan is the clearest case of a country developing a distinct profile of specialization in electronic and mechanical technologies, which is increasingly different from all other countries. Its already high distance indexes increased in the 1980s with almost all countries, confirming the strong specificity of the Japanese technological activities. A special case appears to be the Netherlands, which has moved closer to all the major countries (the US, France, the UK, Germany and Japan), suggesting a major effort at broadening its sectoral distribution of patenting, which is consistent also with previous data on Dutch strengths in some electronic-related technologies (see Chapters 4 to 6).

For the remaining countries, the general trend is towards a growing distance of their specialization profiles, suggesting that most countries develop a sectoral distribution of their patenting activities leading to a more distinct national profile, linked to their own economic and technological characteristics. In this context, the pattern shown by the larger countries appears to be an exception. The strong and growing similarities of the US, the UK and France, and the weaker convergence of Germany, Italy and Switzerland are the result of larger size, making a broader distribution of sectoral activities possible, and are evidence of a specific "path of specialization".

From this complex picture of relative similarities and differences a number of lessons can be drawn. First, the differentiated nature of patenting activity leads to a variety of links between pairs of countries, which are based on shared relative specializations in particular patenting fields. This fragmented pattern of relative similarities makes an overall assessment of the distance between countries difficult, but highlights the different patterns of specialization which are present even within the same country.

Second, the ability of large countries to cover most technology fields with their innovative activities means that size is an important factor in this measure of similarity. This leads to a picture of a "core" composed of the larger countries, which are fairly close to one another, while the smaller ones are scattered around them, closer to one of the larger countries, and highly different from most of the other smaller ones. This result suggests that small countries may benefit from a policy of developing collaboration with other small countries, especially in the cases of clear complementarity. Technology emerges as one of the determinants of the strategies for bilateral agreements often attempted by the governments of small countries (see Freeman and Lundvall, 1988).

Third, while the analysis of individual countries has shown a strong stability of national patterns of specialization in technology (see Cantwell, 1989), comparison of the relative distance between countries reveals some mobility. The differentiated pattern of sectoral patenting and the combined shifts of all countries over time lead to an overall picture of fairly dynamic relative positions of individual countries. In other words, even within the constraints of national technological accumulation, countries do shift their relative positions, seizing or missing the technological opportunities offered by the changing patterns of innovation.

Fourth, we have found that some of the larger countries (the UK and France in particular) have in fact converged to a model of technological specialization similar to the US, while others (Germany, Italy and, more markedly, Japan) have combined the convergence in terms of aggregate resources devoted to science and technology with a growing specialization at the sectoral level. This highlights the different "paths of specialization" which may be followed by countries and points to the importance of the dynamics of integration and competition as a major factor also in international technology.

Finally, these findings on the distances among the profiles of technological specialization raise the question of whether parallel similarities and differences can been found across countries by using such economic indicators as production or trade. The empirical evidence available suggests that this is very much the case and some countries, notably Japan, show a distinct pattern of specialization not only in technology but also in international trade (see Amendola and Perrucci, 1990).

10.3. The Impact of Technological Specialization

The final issue to be addressed is the impact of technological specialization on the performance of advanced countries. A large body of work has documented the importance of technology for economic performance, measured by growth rates (Pavitt and Soete, 1981; Fagerberg, 1988), by international competitiveness (Soete, 1981; 1987; Fagerberg, 1988; Amendola et al., 1991); or by productivity (Nelson, 1981; Baumol et al., 1989).

These studies were carried out at the aggregate level as well as with sectoral analyses. A simple relationship between the countries' general technologcal level and their performance is examined in aggregate studies. Studies at the sectoral level have investigated the relationship between areas of specialization and national performance[3]. While the main influence on economic performance is represented by the quantity of technological resources available and by their sectoral distribution, these are not the only relevant factors. We shall here address two related questions, which derive from the analysis carried out in the previous Chapters, and bring together the aggregate and disaggregate approaches. The two questions can be summarized as follows:

i) Are national performances favourably affected by a high level of specialization in selected fields, or are they better served by a broad distribution of technological efforts across sectors?

ii) Does the composition of technological activities across sectors matter for performance? More specifically, is it advantageous for a country to specialize in fields of growing importance?

In this investigation, the evidence of previous Chapters will be used together with new indicators. Industrial production and patenting in the US will be taken as indicators of national performance as they appear to be appropriate variables to account for the ability of countries to translate their pattern of specialization into overall technological and industrial growth. A country's national system of innovation can be described by many indicators focusing on different aspects and linked by a complex web of relationships. Detailed investigation of the set of links among the many relevant variables and of their impact on national performance is beyond the scope of this book, but an exploratory analysis may help to frame the problem and offer a few suggestions for future research.

We focus here on the data already obtained on the patterns of national specialization and explore which indicator is most relevant and appropriate to summarize the sectoral distribution of a country's technological activities. In this analysis four types of indicators will be considered:

i) Indicators of the size of national technological activities, e.g. cumulative R&D and patents granted in the US, as discussed in Chapters 3 and 8;

ii) Indicators of the innovative dynamism of a country's technological specialization, e.g. the activity in fast growing patent classes, as examined in Chapter 6.

iii) Indicators of the degree of specialization in technological sectors, e.g. the chi square values used in Chapter 8.

iv) Indicators of technological and economic performance, e.g. rates of growth of patents in the US and of industrial production, which will be introduced in this Chapter.

The links between these four types of indicators are summarized here, exploring the common patterns emerging from the experience of advanced countries in the past two decades. Let us first define in greater detail the indicators of the nature and degree of technological specialization to be considered.

i) *Indicators of the size of national technological activities*. The relevance of the size factor has been pointed out throughout this book and need not be repeated here. The two main indicators used, cumulative R&D and patents in the US, provide an adequate description of the aggregate technological effort of countries, and will be used also in this section.

ii) *Indicators of the innovative dynamism of a country's specialization*. In spite of the complexity and variety of the specialization profiles of individual countries, a synthetic indicator (based on disaggregated data by sectors) of the dynamism shown by national specializations in the context of the world innovative activities has been developed. In Chapter 6 the analysis of the fast growing classes of patents in the US (based on the 3-digit IPC classification in 116 subclasses) made it possible to identify the countries which are specialized in areas where inventions grow faster; the indicator used here is the specialization index (TRCA) in the group of classes defined as "fast growing" (see Table 6.3).

iii) *Indicators of the degree of a country's specialization*. The degree of sectoral specialization of individual countries (measured in Chapter 8 by a synthetic index based on disaggregated sectoral data) compares the sectoral percentage distribution of patents granted to an individual country with the sectoral distribution of total patents, providing an indicator of how different (i.e. specialized) the country's profile is from the world's technological activities. The chi square values thus obtained are independent of the number of a country's patents. However, in order to investigate the relationship between the degree of specialization and national performance, a refinement of this index is required. The close (inverse) relationship to size, shown in Chapter 8, needs to be eliminated and an indicator of specialization developed which makes it possible to compare countries of different size. The chi square values were therefore weighted by an indicator of size, i.e. the percentage of patents in the US granted to each country. The result is a weighted degree of specialization which is independent both of the number of a country's patents, and of the effect the scale of national technological activities has on specialization patterns[4].

For a given size of national activity, the higher the chi square value, the greater the weighted index. For a given level of specialization measured by the chi square, the weighted index is greater for the country with larger technological resources, i.e. with a relevant and specialized activity in a broader range of fields. While this is only a preliminary indicator of how active and specialized individual countries are, and many refinements may be suggested, it offers a synthetic indicator which allows a ranking of advanced countries, taking into account both their size and their degree of specialization.

The weighted index was calculated by using the chi squares obtained from data on patents in the US in 1975–81 and in 1982–88, weighted by the share of total patents in the US held by a country in the same periods. Table 10.3 shows the values obtained for the weighted degree of specialization. In both periods the highest indexes are found for Japan, followed by Germany, the EEC aggregate, Switzerland and the US, whose lower value is partly due to the use of patenting in the US domestic market, as pointed out in Chapters 4 and 5. A group of countries with an intermediate level of specialization follows, including Italy, Sweden, the UK, Canada and the Netherlands. A low degree of specialization is found for Belgium, France, Denmark and Spain, while extremely low values are found for Ireland, Portugal and Greece, due to the small scale of their technological activities.

Over the two periods most countries show an increase in their weighted degree of specialization and their ranking changes slightly. The large countries emerge with high values of the index, as they are able to develop a significant specialization in several technological fields, but the marked sectoral specialization of some countries (Japan, Germany, the EEC aggregate and Switzerland) puts them ahead of the US. The ranking of countries draws an already familiar picture. Besides the extremely high values of Japan, we again find a relatively high specialization for Germany, Switzerland and the EEC aggregate. Italy, the Netherlands and the UK follow, while France spreads its technological activities more evenly across sectors; small countries have the lowest values. The results obtained from this new indicator make it possible to compare the level of technological specialization of different countries by putting them on the same footing.

However, it should be stressed that the weighted index of specialization is not

TABLE 10.3

WEIGHTED DEGREE OF SPECIALIZATION IN TECHNOLOGY AND NATIONAL PERFORMANCES

Index of the weighted degree of sectoral specialization in technology and average annual rates of change of patenting in the US and of industrial production 1975-81 and 1981-88

Countries	Weighted degree of specialization		Average annual rates of growth: Patents granted in the US		Average annual rates of growth: Industrial production	
	1975-81	1982-88	1975-81	1981-88	1975-81	1981-88
United States	0.58	0.72	-2.88%	0.49%	3.37%	3.22%
Japan	1.40	2.62	4.76%	9.78%	4.75%	3.84%
EEC	0.74	0.87	-0.49	2.14	2.15	2.24%
W. Germany	0.73	0.95	0.64%	2.18%	2.62%	2.12%
France	0.13	0.13	-1.33%	2.79%	2.08%	2.14%
Un.Kingdom	0.24	0.24	-3.18%	0.48%	0.66%	3.20%
Italy	0.26	0.32	3.17%	2.83%	3.45%	2.12%
Netherlands	0.24	0.22	0.56%	3.26%	2.00%	1.78%
Belgium	0.13	0.14	-0.95%	1.41%	2.04%	2.52%
Denmark	0.06	0.07	-2.19%	2.37%	0.61%	2.46%
Spain	0.06	0.06	-8.09%	11.52%	1.40%	2.94%
Ireland	0.02	0.01	2.25%	13.96%	3.79%	1.71%
Portugal	0.01	0.01	-10.20%	4.52%	3.99%	0.33%
Greece	0.01	0.01	-9.87%	7.88%	3.90%	0.30%
Switzerland	0.74	0.60	-2.67%	0.04%	1.62%	1.91%
Sweden	0.32	0.28	-2.94%	0.27%	-0.73%	2.83%
Canada	0.23	0.26	-2.20%	3.62%	3.89%	2.59%

Source: CNR-ISRDS elaboration on CHI Research patent data and OECD, Main Science & Technology Indicators data, april 1990.
Note: The weighted degree of specialization is obtained by weighting the degrees of specialization measured by chi square values for patents granted in the US, shown in table 8.1, with the percentage of US patents held by each country.

an indicator of the "quality" of the specialization profile of individual countries, and therefore should not be interpreted as suggesting "how good" a country's position is. Rather, it indicates how relevant and distinctly specialized a country's technological activities are. These characteristics describe a particular aspect of national activities in technology and need to be further qualified, for instance with the information offered by the above indicator of innovative dynamism of the fields of national specialization.

iv) *Indicators of national performance.* Two new indicators will be used here to describe the economic and technological performance of each country. Firstly, the average annual rate of change of patents granted in the US within each period will be used as an indicator of technological activity in the single largest market of the world. While patenting in the US is not an entirely satisfactory indicator of the overall growth of a country's innovative efforts, it provides some relevant data and makes it possible to compare the technological positions of foreign countries in the US market. Secondly, the average annual rate of change in real terms of the value of industrial production within each period will be considered as an indicator of economic performance in the industrial sector. Industry is the part of the economy where technological innovation is expected to have the most direct and relevant effects. While offering only a preliminary and limited description of national performance, these two variables appear appropriate to assess the impact of different degrees of national specialization in technological sectors.

The rates of change of both patenting and industrial production for most countries show substantial changes over the two periods and in examining national performance particular attention should be paid to the great influence of the economic cycle on industrial and also technological indicators of growth. The two periods considered start with major downturns of the economy, in 1975 and in 1981–82, and appear to pinpoint the two most recent cycles. However, this periodization also has its limitations, as the timing of the cycle shown by individual countries may be different, and data on rates of growth are highly volatile. All these points need to be borne in mind in evaluating the association between technological specialization and national performance. Table 10.3 presents the average annual growth rates of these variables for all countries, calculated from 1975 to 1981 for the first period, and from 1981 to 1988 for the second.

The Links between Technology and Performance Indicators

A review of the set of relationships existing between the four groups of indicators – size, innovative dynamism, degree of specialization and performance – provides a picture of the common patterns of technological activities in advanced countries. A preliminary overview of these links was obtained by examining correlation coefficients between each pair of the variables defined above, calculated for the two periods considered (1975–81 and 1982–88). The 13 major countries were considered here, leaving out Ireland, Portugal and Greece, the three smallest countries, whose data are generally less significant and reliable[5].

Firstly a set of relationships within the structure of technological activities was considered. The scale of national efforts, the degree of specialization and national innovative dynamism were found to be largely unrelated in both periods, suggesting that they represent different aspects of national technological systems with separate roles and diverging dynamics. Secondly, the impact of each aspect of the national systems of innovation on countries' performances was examined. The three sets of variables described above were related to a technological indicator (the rate of growth of patents in the US), and to an economic indicator (the rate of growth of industrial production).

The size of innovative activities of nations showed no relation to the growth of patents, and was weakly associated to that of industrial production (for the latter, correlation coefficients are in both periods about 0.40). Size is therefore not a determinant factor for the growth of national technological activities measured by patents in the US, while the industrial performance of small countries has often shown lower growth rates than larger countries.

No relation emerged in either period between a country's innovative dynamism and its performance. Countries specialized in technological fields where innovations grow faster did not show higher than average increases either in overall patenting activity or in industrial production. It is likely that the fast growing areas of innovation are rather small fields with, at first, little impact on total national patenting activities. Moreover, such fields may not correspond to already established industries and markets, which may develop in the future, as innovations are translated into new products and processes. A significant time lag may therefore exist between a country's early advantage in new innovative fields and a visible improvement of its overall technological and industrial performance. Furthermore, a country's technological and industrial growth is affected by the sectors of declining national activity as much as by those showing rapid growth. A rapid decline in traditional sectors where national activities are concentrated may damage its performance in spite of a strong advantage in innovative areas of still limited industrial importance.

Conversely, a clear positive correlation emerged in both periods between the degree of specialization and increase in patenting (coefficients equal to 0.55 and 0.39) and industrial production (coefficents of about 0.53). Countries with a relevant and specialized activity in particular sectors tend to perform better in both technological and industrial terms than countries which spread their activities more evenly across fields. The degree of specialization (weighted with the countries' size) therefore emerges as a technological variable with a relevant impact on national performance. This confirms our expectation that a high degree of specialization contributes positively to the growth of technological and industrial activity. In other words, it appears that there is a general advantage in being specialized; the advantages in innovative activities derived by economies of scale and economies of scope seem to be, at the national level, more important that the choice of "good" sectors. In order to explore this relationship in greater detail and to investigate the position of individual countries, a more detailed analysis is carried out.

Figure 10.1 shows the relationships between degree of specialization and rates of growth of patents in the US in the two periods considered. Figure 10.2 shows the same for the rates of growth of industrial production. The regression lines are drawn to illustrate the general pattern, thus making it possibile to identify the relative position of individual countries and their changes over time. Full information on the regression equations is provided by Table 10.4.

The degree of technological specialization is closely related to the growth of patenting in the US in the 1975–81 period (R square equal to 0.30), and shows a weaker association in the second period 1982–88 (R square equal to 0.15)[6].

Figure 10.1 shows that in both periods Japan has the highest values of the degree of specialization and the growth rate of patents in the US. The EEC aggregate[7], Germany, Switzerland and the US are the areas with the next highest levels of specialization, but they tend to show much lower growth rates of patenting. Among the remaining countries with lower degrees of specialization, Italy, the Netherlands and France show a higher than average growth of patents in the US in both periods, while Sweden and the UK show a lower growth rate in both periods.

Looking at the evolution of the distribution of countries over time, we find that the EEC aggregate, Germany and the US follow the same trajectory of growth along both variables traced by Japan at a distance. Most other countries show a

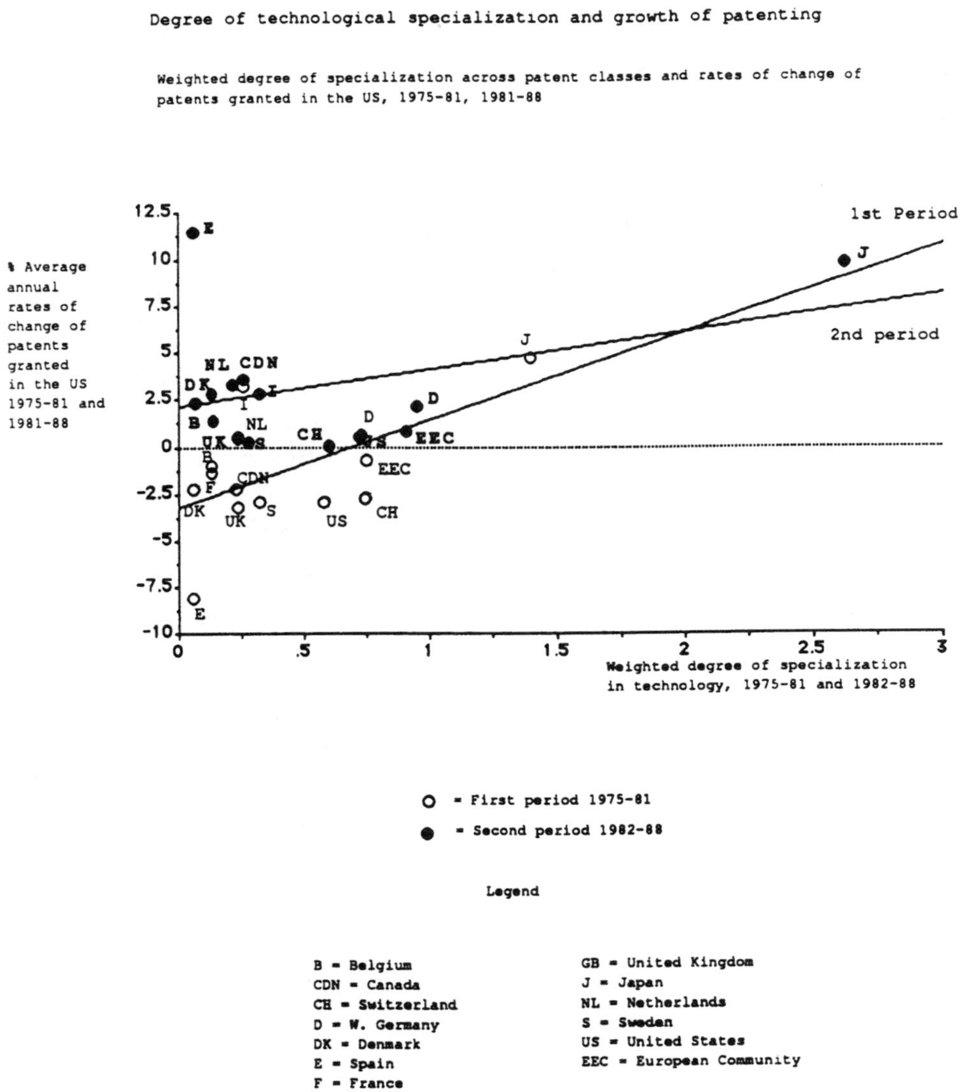

Fig. 10.1.

marked increase in their rates of change in patenting, often combined with small increases in their degree of specialization. Spain appears to be an extreme case, due also to the country's institutional changes, thus confusing the overall pattern a little.

The degree of technological specialization is also closely related to the growth of industrial production (R squares equal to 0.28 and 0.30) and Fig. 10.2 shows the distribution of countries in the two periods considered. Here again Japan has the highest values in both periods, followed by the US, Germany and the EEC aggregate, with high technological specialization and high growth rates of industrial

Degree of technological specialization and growth of industrial production

Weighted degree of specialization across patent classes and rates of change of industrial production in real terms, 1975-81, 1981-88

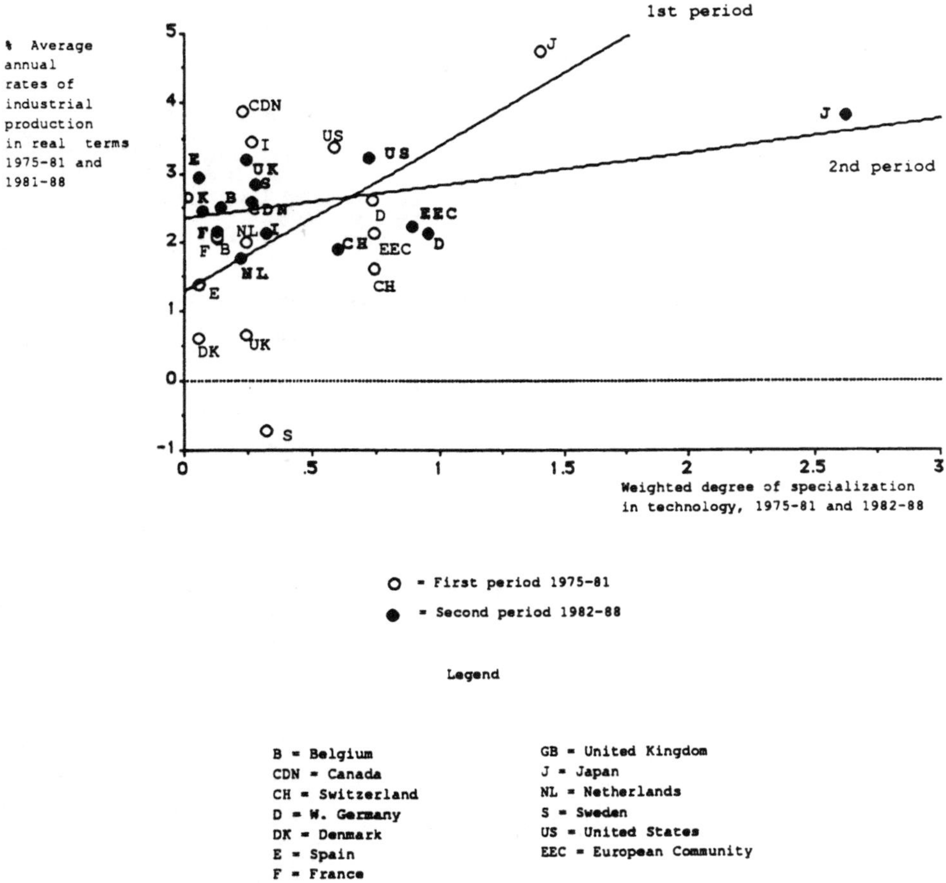

Fig. 10.2.

production (except for Germany in the second period). Among the countries with lower specialization levels, a strong instability over time is found, with those experiencing the fastest growth of industrial production in the first period showing lower than average values in the second and vice versa. Over the two periods no clear pattern of change can be found in the distribution.

These findings show that the greater the ability of a country to develop a comparative advantage in selected sectors, the better its performance generally is in technological and industrial terms. Some countries, Japan in particular, show very high values for all these variables in both periods. The UK tends to show, on the opposite, very low values for most variables. The strong technological specializa-

TABLE 10.4.

THE RELATIONSHIPS BETWEEN TECHNOLOGICAL SPECIALIZATION AND NATIONAL PERFORMANCES

List of variables:

W1: Weighted degree of technological specialization, 1975-81.
W2: Weighted degree of technological specialization, 1982-88.
VPAT1: Percent average annual rates of change of patents granted in the US, 1975-81
VPAT2: Percent average annual rates of change of patents granted in the US, 1981-88
VPIN1: Percent average annual rates of change of industrial production in real terms, 1975-81
VPIN2: Percent average annual rates of change of industrial production in real terms, 1981-88

1. Degree of specialization and rates of growth of patents in the US, 1975-81
   VPAT1 = -3.152 + 4.635 W1
   t value of beta = 2.211
   R square = 0.308; Adjusted R square = 0.245

2. Degree of specialization and rates of growth of patents in the US, 1981-88
   VPAT2 = 2.132 + 2.014 W2
   t value of beta = 1.408
   R square = 0.153; Adjusted R square = 0.076

3. Degree of specialization and rates of growth of industrial production, 1975-81
   VPIN1 = 1.318 + 2.077 W1
   t value of beta = 2.072
   R square = 0.281; Adjusted R square = 0.215

4. Degree of specialization and rates of growth of industrial production, 1981-88
   VPIN2 = 2.346 + 0.477 W2
   t value of beta = 2.173
   R square = 0.300; Adjusted R square = 0.237

tion of Germany has a greater impact on the growth of patenting than on that of industrial production, and the opposite is found for the US. Italy and France, with lower levels of specialization, follow the pattern of Germany, while strongly specialized Switzerland shows a poorer performance in both variables. The position of the EEC aggregate is generally very close to the pattern of Germany.

It should be noted that the performance indicators themselves do change substantially over the two periods considered: the countries with strong growth rates between 1975 and 1981 often slow down in the following years, and vice versa. It is all the more significant that the relationship with the degree of technological specialization remains evident over time, although it is weaker for the growth of patenting in the second period.

The positive association between the weighted degree of specialization in technology and the performance indicators found here obviously does not mean that the growth of patenting activity and of industrial production can be explained by the different levels of national specialization alone. As already pointed out, national performances have to be explained mainly by the technological intensity of the countries' economies and by the characteristics of their national systems of innovation. The other factors that should be considered here include the type of sectoral specialization shown by individual countries, the expected time lags in the transformation of technological innovations into advantages in industrial production, the influence of business cycles, the links between sectoral strengths in technology and in production, and the level of integration among technological activities in different sectors.

Some information has however emerged from this analysis. Among the factors contributing to better national performance we found that the degree of technological specialization does play a significant role. This result contributes new evidence to current literature on technological change. The presence of specialized technological activities has a generally positive association with better technological and industrial performance in advanced countries. This is a finding which has clear implications for policy, since it suggests that countries have in any case an incentive to specialize, regardless of the sectors selected. In other words, it is better to be specialized even in the "wrong" sectors than not to be specialized at all.

Notes

1. Data on patents granted in the US are used here and particular caution is therefore needed in interpreting the results for the US, due to "domestic market effect" pointed out in Chapter 4 (see Table 4.5).
2. The maximum value of the distance indicator is found in the very special case when two countries concentrate all their activity in two different sectors which have the lowest shares of patents in the world distribution. The lowest these shares, the greater the maximum, which also depends on the number of classes of the distribution. In order to avoid erratic values, the average of the five sectors with the lowest shares of the world distribution has been considered in calculating the maximum. While such a standardization does not affect the relative position of countries in a given period, it is required in order to make comparisons over time, as is done in Tables 10.1 and 10.2. We are indebted to Maurizio Vichi for his advice in developing this index.
3. They have generally used data on the sole *production* of innovations rather than the total technological intensity of each sector, which ideally should include bot the innovations produced and those used in the sector.
4. We are indebted to Maurizio Vichi for the development of this index.
5. The same analysis has been carried out also for all 16 countries, finding broadly similar patterns of association among the variables, but less stability over the two periods considered.
6. This lower R square is due to the extremely high growth rate of patents shown by Spain (which is largely an institutional artifact); if Spain is removed from the analysis, the R square value for the equation is 0.58.
7. The values for the EEC aggregate are shown in the graph, but obviously have not been included in the calculations of either the correlation index, or the regression line.

CHAPTER 11

Conclusions

Technology is widely regarded as a key factor contributing to economic growth and international competitiveness. The extent of its contribution depends on the way countries are able to develop, select, acquire, and disseminate know how through their national systems of innovation. This book has focused on the sectoral structure of national technological activities and on the process of specialization. Our analysis should be seen in the context of the debate on the transformation of advanced economies, on their convergence in terms of aggregate indicators and industrial structures, and on their changing relative positions. It documents the convergence of advanced countries in terms of their S&T intensity, and the growing concentration of national efforts in selected fields of international excellence.

Specialization is the key concept in this book, and a restatement now may be appropriate. First, it should be recalled that we have examined specialization patterns at the national level, rather than at the level of industries or firms. Although countries are challenged by a process of internationalization affecting science and technology, as well as the economy and society as a whole, our research supports the view that capacities are often country-specific. The behaviour of firms, including those of multinational corporations, does in fact take into account the relative strengths and weaknesses of individual countries in innovative activities.

Secondly, specialization has been considered in relation to a fairly disaggregated classification of technologies and industries. This analysis complements the widely available evidence on aggregate national S&T activities based on standard indicators, and provides a more complex and detailed picture of individual countries, using much broader empirical evidence. In other words, the sectoral analysis offers qualification and specification of aggregate patterns. Based on patenting and bibliometric data, the indicators employed have provided internationally comparable information on a large number of industrial countries over a wide spectrum of science and technology fields.

Thirdly, specialization refers, by definition, to what countries do or do not do in comparison with other countries. The analysis focuses on the relative positions of nations, and emphasizes the differences in, rather than similarities between, their innovative activities. This should not lead us to overlook the absolute changes in the level and composition of S&T efforts and, in particular, the growing importance of certain broad fields (e.g. electronics) which is common to most countries, as documented by several case studies in the areas of new technologies.

The results of the analysis carried out in this book can be summarized as follows:

1) The advanced economies are becoming more knowledge intensive, as indicated by the large expansion of science and technology activities. While in the past two decades most OECD countries have experienced rapid growth, by the end of the 1980s signs of a slow down of domestic efforts (measured by R&D expenditure and number of patents) could be found in some countries, as documented in Chapter 3. A steady growth in the major national inputs for innovation, such as R&D funds, cannot be assumed for the future, while other sources of innovation can be expected to increase in importance. These include activities carried out outside R&D laboratories (in particular informal innovative efforts in the production departments of firms) and innovations acquired from abroad.

2) Cross-border flows of technology (including patents extended abroad, technological balance of payments transactions, cooperation in research projects and coauthorship of scientific papers) have continued increasing rapidly, and now represent a major source of innovation for all countries, from the small open economies of Europe to the large US market. The internationalization of technological activities is now a key aspect of an increasingly integrated global technology market.

3) In the activity of firms, the pressure of international and domestic competition forces individual companies to expand their know how. The R&D activities carried out within the firm tend to focus on carefully selected research priorities, and increasing attention is devoted to new ways of acquiring know how from outside. R&D cooperation between firms has set a major new pattern leading to a variety of agreements, from joint research projects to cross-licensing, from investment in small innovative firms to joint participation in government-sponsored R&D programmes. Firms are increasingly exploiting their innovations in international markets, as indicated by the growth in the number of foreign and external patent applications. This company strategy aiming at appropriating returns on innovative efforts in the global market is leading to changes in the distribution of technological activities in individual countries.

4) The long-term increase in national R&D resources and international technology flows has led to a major change from the post-war US technological leadership to a remarkable convergence of the innovative intensity of advanced countries. Most OECD countries have consistently shown a higher growth rate of S&T activities than the US, and a group of countries, including Japan, Germany, Sweden and Switzerland, show R&D intensity comparable with that of the US. Moreover, the US is characterized by its focus on military technology, and its civilian R&D as a share of GDP is lower than in several OECD countries, as shown in Chapter 3. The result is that the "technology gap" between the US and the rest of the world now appears to have closed. However, in Japan the growth rate of S&T activities in both input and output indicators has been substantially higher than for the aggregate of the European countries.

5) Along with this growth and convergence of aggregate S&T activites in advanced countries, differing areas of national strength and weakness in science and technology activities have been emerging at the sectoral level. The empirical analysis of the sectoral distribution of patenting activities in Chapters 4 and 5, and of

scientific literature in Chapter 7 has provided a detailed picture of the structure of S&T activities in each country. We have shown that national systems of innovation differ not only in their nature and ways of introducing innovations, as argued in the literature, but also in the measurable outcomes of their activities at the sectoral level.

Such striking diversity of national profiles in science and technology was not an obvious outcome. Several case studies on technological change in individual industries have shown that most countries devote a growing share of their resources to selected new technologies, suggesting a common shift towards "advanced" sectors. Our findings suggest, however, that within such a common technological transition, the relative position of countries in disaggregated fields is still marked by growing divergence and specialization.

In other words, the "catching up" process has not driven the European countries and Japan to follow and replicate the US pattern of sectoral technological strengths. On the contrary, each country has developed a distinct model of specialization, concentrating its efforts in particular fields where world class capacities have often been developed. Most countries therefore appear now present at the technological frontier at least in the selected fields of their major specialization. European countries maintain a rich variety in their areas of strength, which may be an asset in the future evolution of the international technology system.

6) Qualification of the areas of national technological specialization has been provided in Chapter 6 with the analysis of groups of patent classes ranked according to their rates of change, thus pinpointing the fields of greater innovative dynamism. Even greater differences have been found in the position of individual countries across such groups. Although the European countries have performed rather well in terms of their aggregate rate of growth in technology, a major weakness has emerged in the fields with more intense technological change. The sectors where most EEC countries are specialized tend to fall disproportionately into classes where patenting activity is slow-growing or declining, casting serious doubts on Europe's prospects of further improving its technological position. Japan, on the other hand, is the only country concentrating its technological strengths in the fields showing faster growth of patents. However, specialization in fast growing sectors does not seem to be generally associated with better economic or technological performance, as revealed by the analysis in Chapter 10. The fast growing patenting fields tend to be of smaller size and to focus on developments at the technological frontier, while the bulk of industrial activities is in fields with more mature technologies. The extent to which such new technologies will affect national performances is highly uncertain, but Europe's weakness in fast growing areas of innovation may jeopardize its long term competitiveness.

7) In this book great attention is devoted to the empirical analysis of the degree of national specialization in technology and science. We find that most countries have increased their degree of specialization in technological activities, while the opposite pattern of declining specialization has emerged in scientific efforts, as documented in Chapter 8. The evidence on technology suggests that nations are concentrating their efforts in the few areas where they have an established advantage. These fields differ from country to country, and the resulting pattern

is one of growing specialization. For scientific activity we find the opposite trend, leading to a broader distribution of efforts across fields. The contrasting science and technology patterns result from the different nature of the two activities, and from institutional differences in the innovative systems of advanced countries.

Increasing international integration of technological activities and growing national specialization in selected fields emerge from this analysis as the two aspects of the same process. Global competition leads individual countries to focus on their areas of international excellence, consolidating the pattern of national comparative advantages, while increasing reliance on the acquisition of increasingly accessible foreign-developed know-how for the fields of national weakness. This process has opened the way to rapid productivity growth in several countries, resulting in the convergence of aggregate performance indicators. At the same time the sectors of national activity emerging from the selection process have become increasingly differentiated. This outcome is affected by the international technology strategies of large firms, deciding where to carry out research and where to disseminate their technology. However, the comparative advantages offered by the different national systems of innovation do undoubtedly remain a major force influencing the choices of large multinational firms.

The result of these processes is that the world technological landscape is now characterized by a more fragmented pattern of relative national advantages in selected technology fields. No country appears able to show (or aim at) uniformly strong performance in all sectors. The case of Japan, the most succesful country over a whole range of technological efforts, shows that the best performance has been associated with the deliberate strategy of focusing on selected areas.

8) The size of countries has obviously emerged as a constraint on the development of domestic capacities and smaller countries are under greater pressure to integrate their innovative activities internationally. Our analysis has shown that the degree of specialization in technology is inversely related to the size of the country. While large countries distribute their innovations more uniformly across most sectors, small countries are forced to higher levels of specialization. This result has long been seen in the case of trade and is here empirically documented on the basis of technological indicators. The same relationship holds also for indicators of scientific activity, although somewhat less marked.

9) An overview of the national specialization patterns is provided by analysis of the distance (measured with a specially devised index) between the profiles of technological activities of pairs of countries. Using this method we described the relative positions of individual countries and their similarities and differences in Chapter 10. Larger countries show stronger similarities as they are able to cover more fields of technological effort, while smaller countries appear more distant, and show widely differing specialization profiles. They come under pressure to concentrate in niches of international excellence, but the choice of particular sectors does not appear seriously constrained by smaller size. For almost all pairs of countries the distance increases over time, confirming that nations develop increasingly distinct sectoral distributions of innovative activities.

10) The process of technological specialization is the result of different trends across the sectors. In examining the country to country distribution of technological

activities in each area, we find that most production-intensive fields show innovative efforts tending to concentrate in a few countries, while only the more pervasive and generic technologies show increasingly uniform international distribution, as documented in Chapter 9. In other words, innovative activities within most fields appear to be taking place in a decreasing number of countries, which in turn emerge with a growing specialization in such fields, resulting in more markedly polarized country to country distribution of technological efforts in individual sectors. The distinction between the few fields showing more uniform distribution and the majority of sectors undergoing greater international polarization should be borne in mind when addressing industry case studies.

11) The final question addressed in this book concerns the impact of technological specialization on national performance. A higher degree of specialization (weighted in order to take into account the size of the country) is associated with a more rapid growth of patenting activities and of industrial production. Countries with a strong, specialized innovative activity in seleceted sectors tend to show better industrial and technological performance than countries with smaller and more uniformly distributed efforts. There seems to be a specific advantage in a higher degree of specialization in technological fields, associated with the economies of scale and scope made possible at the national level. This advantage emerges regardless of the particular sectors in which individual countries concentrate their efforts; in other words, for advanced countries being specialized appears to be even more important than choosing the "right" fields.

These findings suggest certain integrations and new departures for research in technology policy and the economics of production and trade.

First, the role played by the degree of technological specialization as a determinant of innovative and economic performances needs to be explored further. So far studies on the relationship between technology and performance have either focused on aggregate variables (e.g. the national levels of technological capacities and the rates of change of performance indicators), or on the relationship between technology and performance across sectors (see Pavitt and Soete (1981), Soete (1987), Fagerberg (1987, 1988), Amendola, Dosi and Papagni (1991)). We have shown that a higher degree of sectoral specialization can also have an impact on national performance. Additional research is needed in this direction, with specification of the impact different patterns of specialization have, including an appropriate typology of sectoral specializations and of the vertical and horizontal linkages existing between the fields of national strength.

Secondly, the patterns identified for technological specialization should be compared to those emerging from other economic indicators such as production and trade, which are equally affected by the process of internationalization (for an attempt, see Amendola, Guerrieri and Padoan, 1991). A growing degree of specialization can also be expected for these variables, which may show greater concentration of national activities in highly selected fields of national comparative advantage. An even stronger process of specialization could be expected within the more knowledge-intensive sectors, as for instance in the case of high technology products.

Thirdly, while increasing specialization can be expected for supply indicators, the opposite process may occur on the demand side. The sectoral polarization of international technological and economic activities could be associated with an increasing convergence in the patterns of final consumption, with different patterns emerging in the demand for investment and intermediate goods. Exploration of this hypothesis may show whether countries are specializing in increasingly varied production activities while moving to closer final consumption patterns, a divergence which can be bridged only by increasing international trade including both intra-industry and inter-industry flows.

Policy Implications

The findings of this book may offer new perspectives for national technology policies. The detailed analysis of strengths and weaknesses of the various countries in science and technology, of their similarities, differences and evolution over time can contribute to a better assessment of the prospects for national systems of innovation. In terms of policy options, two major opposing pressures appear to shape national choices. On the one hand the importance of technological accumulation and the persistence in time of some sectoral specializations suggest policies of defending and building up consolidated advantages in the face of foreign competition. Better economic and technological performance is more likely to be achieved by persisting in the established sectors of strength, even if they happen to be in low technology fields, rather than by shifting to fields of scant national competence.

On the other hand, there is the pressure to pursue the opportunities opened by technological advances in new fields, challenging the advantages of the other countries. The changing relative position of individual countries in innovative activities can be understood as a process of growing endeavour and specialization in technology, with countries facing challenges to their traditional advantages and seeking new advantages in other fields. The importance of innovating in new technological fields is obvious, but such efforts are risky and require a considerable amount of resources and favourable institutional conditions. They also have to meet more intense competition in global markets while their payoffs may take years to materialize. In allocating limited national resources to science and technology, difficult choices have to be made at the sectoral level.

Careful targeting of technology is required at the industry level, taking into account the different dynamics shown by individual sectors. For instance, opposing patterns have emerged between generic and pervasive technologies on the one hand, where the international distribution of activities is more even and, on the other hand, the fields showing growing international competition and specialization, based on economies of scale and cumulative advantages. Increase in national specialization is also associated with the growth of international cooperation among firms, research institutions and governments. These collaborative arrangements do not change the competitive nature of relations between firms, but have emerged as key elements in the technology policy of advanced countries. Identifying complementarities between the different specializations developed by various agents is likely to be an increasingly important aspect of national technology policy. The experience of the EEC high technology programmes is an important asset in the search for cooperative innovative activities, and it may deserve to be extended to other fields and to a wider range of agents in the business and research communities.

While greater resources and a more specialized pattern of technological activites appear to be associated with better national performance, several economic and institutional conditions play an equally important role in the successful development of a country's innovations. In this context, countries will find their resourcefulness increasingly challenged by a technology system which has grown more international and cooperative in orientation, specialized in selected sectors, and constrained by intense international competition. In the 1990s, the outlook for technological advance is clouded by signs of a slow-down in the R&D efforts of many countries, by stagnant patterns of domestic patenting and by uncertainty over the priorities of technological efforts. The economic recession which has hit a number of countries casts yet another shadow over the future, and stagnating R&D efforts certainly do not contribute to renewed growth of advanced economies. Given the present context, the processes of technological change and specialization investigated in this book may suggest more sharply focused and appropriate policies for innovation and new directions for national efforts in science and technology. In turn, together with very many other economic and social factors, these may contribute to improving the prospects for advanced countries in the new century.

References

Achilladelis, B., A. Schwarzkopf and M. Cines (1987) *A Study of Innovation in the Pesticide Industry: Analysis of the Innovation Record of an Industrial Sector*, in Freeman, 1987b.

Acs, Z.I. and D.B. Audretsch (1990) *Innovation and Small Firms*, The MIT Press, Cambridge, Mass.

Amendola, G. and A. Perrucci (1988) *Commercio Internazionale e Tecnologia: Aspetti Teorici e Problemi di Ricerca Empirica*, ENEA, Rome.

Amendola, G. and A. Perrucci (1990) *Technology and Competitiveness. Specialization and Competitiveness of Italian Industry in High-Technology Products: A New Approach*, OECD Conference, Paris, 24–27 June.

Amendola, G., P. Guerrieri and P.C. Padoan (1991) *International Patterns of Technological Accumulation and Trade*, Paper for the EARIE Conference, Ferrara, 1–3 September.

Amendola, G., G. Dosi and E. Papagni (1991) *The Determinants of International Competitiveness*, ENEA, Rome, mimeo.

Antonelli, C. and L. Pennacchi (eds.) (1989) *Politiche dell'Innovazione e Sfida Europea*, Angeli, Milan.

Arcangeli, F., P. David and G. Dosi (eds.) (1989) *The Diffusion of New Technology: Modern Patterns in Introducing and Adopting Innovations*, Oxford University Press, London.

Archibugi, D. (1986) *Paradigms and Revolutions: From Science to Technology?*, Paper Presented at the EASST 4th Annual Meeting, October 1986, Strasbourg.

Archibugi, D. (1988a) The inter-industry distribution of technological capabilities. A case study in the application of the Italian patenting in the USA, *Technovation*, vol. 7, n. 3, pp. 259–274.

Archibugi, D. (1988b) In search of a useful measure of technological innovation, *Technological Forecasting and Social Change*, vol. 34, n. 3, pp. 253–277.

Archibugi, D. (1989) *The Sectoral Structure of Innovative Activities in Italy. Results and Methodology*, Thesis Submitted for a Ph. D. Degree, University of Sussex, Brighton.

Archibugi, D., S. Cesaratto and G. Sirilli (1987) Innovative activity, R&D and patenting: the evidence of the survey on innovation diffusion in Italy, *Science Technology Industry Review*, vol. 1, n. 2, pp. 135–150.

Archibugi, D. and R. Malaman (1991) Il brevetto come strumento di appropriazione dell'attivita' inventiva ed innovativa: l'insegnamento delle indagini empiriche, in Malaman, 1991.

Archibugi, D. and M. Pianta (1989) *The Technological Specialization of Advanced Countries*, Commission of the European Communities, Ad Interim Report, Brussels.

Archibugi, D. and E. Santarelli (1989) Tecnologia e struttura del commercio internazionale: la posizione dell'italia, *Ricerche Economiche*, vol. 43, n. 4, pp. 427–455.

Audretsch, D. and H. Yamawaki (1988) R&D rivalry, industrial policy and US – Japanese trade, *The Review of Economics and Statistics*, vol. 70, n. 3, pp. 438–447.

Balassa, B. (1965) Trade liberalization and trade in manufactures among the industrial countries, *American Economic Review*, vol. 56, pp. 466–473.
Basberg, B. (1983) Foreign patenting in the USA as technology indicator: the case for Norway, *Research Policy*, vol. 12, n. 4, pp. 227–237.
Basberg, B. (1987) Patents and the measurement of technological change: a survey of the literature, *Research Policy*, vol. 16, n. 2–4, pp. 131–141.
Baumol, W.J., B.S.A. Blackman and E.N. Wolff (1989) *Productivity and American Leadership: The Long View*, MIT Press, Cambridge, Mass.
Bertin, G. and S. Wyatt (1986) *Multinationales et Propriété Industrielle. Le Controle de la Technologie Mondiale*, Presses Universitaires de France, Paris.
Bound, J. et al. (1984) *Who Does R&D and Who Patents?* in Griliches, 1984.
Braun, T., W. Glanzel and A. Schubert (1989) Assessing assessments of British science: some facts and figures to accept or decline, *Scientometrics*, vol. 15, n. 3–4, pp. 165–170.
Cantwell, J. (1989) *Technological Innovation and Multinational Corporations*, Basil Blackwell, Oxford.
Cantwell, J. and C. Hodson (1990) *The Internationalization of Technological Activity and British Competitiveness: A Review of some New Evidence*, Department of Economics, University of Reading, mimeo.
Cantwell, J. (1991) *The Technological Competence Theory of International Production and its Implications*, Department of Economics, University of Reading, mimeo.
Carpenter, M.P., F. Narin and P. Woolf (1981) Citation rates to technologically important papers, *World Patent Information*, vol. 3, n. 4, pp. 160–163.
CER (1988) *L'Attivita' Innovativa in Italia: I Brevetti nell'Industria*, Rome.
Chesnais, F. (1988) *Multinational Enterprises and the International Diffusion of Technology*, in Dosi et al., 1988.
Chi Research – Computer Horizons (1988) *An Analysis of Technological Strength among Japan's Major Corporations*, Venture Economics.
Chi Research – Computer Horizons (1989a) *Technological Activity and Impact Indicators Database*, Tape Data 1975 through 1988, Data User Manual, Haddon Heights.
Chi Research – Computer Horizons (1989b) *Science Literature Indicators SP 2 Subfield Citation Tape 1973–1984 and 1981–1986*, Data User Manual, Haddon Heights.
Coward, H.R. and J.J. Franklin (1989) Identifying the science-technology interface: matching patent data to a bibliometric model, *Science, Technology and Human Values*. vol. 14, n. 1, pp. 50–77.
Dasgupta, P. and P. Stoneman (eds.) (1987) *Economic Policy and Technological Performance*, Cambridge University Press, Cambridge.
Denison, E.F. (1967) *Why Growth Rates Differ*, Brookings Institution, Washington, DC.
Dollar, D. and E.N. Wolff (1988) Convergence of industry labor productivity among advanced economies, 1963–82, *The Review of Economics and Statistics*, vol. 70, n. 4, pp. 549–558.
Dollar, D. and E.N. Wolff (1989) *Trade Patterns and Productivity Convergence among Industrial Countries, 1970–86*, mimeo.
Dosi, G. (1988) Sources, procedures and microeconomic effects of innovation, *Journal of Economic Literature*, vol. 26, n. 3, pp. 1120–1171.
Dosi, G., C. Freeman, R. Nelson, G. Silverberg, L. Soete (eds.) (1988) *Technical Change and Economic Theory*, Frances Pinter, London.
Dunning, J.H. (1988) *Multinationals, Technology and Competitiveness*, Unwin Hyman, London.
Engelsman, E.C. and A.F.J. Van Raan (1990) *The Netherlands in Modern Technology: A Patent-Based Assessment*, Centre for Science and Technology Studies, Leiden.
European Patent Office, *Annual Report*, various years, Munich.
Evenson, R.E. (1984) *International Invention: Implications for Technology Market Analysis*, in Griliches, 1984.
Evenson, R.E. (1989) *Patent Data: Evidence for Declining R&D Potency?*, Yale University, New Heaven, mimeo.
Fagerberg, J. (1987) *A Technology Gap Approach to Why Growth Rates Differ*, in Freeman, 1987b.
Fagerberg, J. (1988) International competitiveness, *Economic Journal*, vol. 98, pp. 355–374.
Faux, J. (1988) The austerity trap and the growth alternative, *World Policy Journal*, vol. 5, n. 3, pp.

367–413.
Finan, W. et al. (1986) *The US Trade Position in High Technology: 1980–1986*, A Report Prepared for the Joint Economic Committee of the US Congress, Washington D.C.
Flamm, K. (1984) Technology policy in international perspective, in *Policies for Industrial Growth in a Competitive World*, Joint Economic Committee, U.S. Congress, 27 April, 1984.
Frame, J.D. and F. Narin (1988) The national self-preoccupation of American scientists: an empirical view, *Research Policy*, vol. 17, n. 4, pp. 203–212.
Freeman, C. (1982) *The Economics of Industrial Innovation*, Frances Pinter, London.
Freeman, C. (1987a) *Technology Policy and Economic Performance. Lessons from Japan*, Frances Pinter, London.
Freeman, C. (ed.) (1987b) *Output Measurement in Science and Technology*, North-Holland, Amsterdam.
Freeman, C. and B.A. Lundvall (eds.) (1988) *Small Countries Facing the Technological Revolution*, Frances Pinter, London.
Freeman, C. and L. Soete (eds.) (1987) *Full Employment and Technical Change*, Basil Blackwell, Oxford.
Garfield, E. (1979) *Citation Indexing; Its Theory and Application in Science, Technology, and Humanities*, Wiley, New York.
Giersch, H. (ed.) (1981) *Emerging Technologies: Consequences for Economic Growth, Structural Change and Employment*, J.C.B. Mohr Paul Siebeck, Tübingen.
Granstrand, O. (1982) *Technology, Management and Markets*, Frances Pinter, London.
Grevink, H. and H. Kronz (1985) *Trends in Patent Filing Activities in the EEC*, Information Management, Commission of the European Communities, Luxembourg.
Grevink, H. and H. Kronz (1986) *Trends in Patent Filing Activities in the USA*, Information Management, Commission of the European Communities, Luxembourg.
Griffith, B.C., H.G. Small, J.A. Stonehill and S. Dey (1974) The structure of scientific literature II: toward a macro and micro structure for science, *Science Studies*, vol. 4, n. 4, pp. 339–365.
Griliches, Z. (ed.) (1984) *R&D, Patents and Productivity*, University of Chicago Press, Chicago.
Griliches, Z., A. Pakes, B. Hall (1987) *The Value of Patents as Indicators of Innovative Activity*, in Dasgupta-Stoneman, 1987.
Griliches, Z. (1990) Patent statistics as economic indicators: a survey, *Journal of Economic Literature*, vol. 28, n. 4, pp. 1661–1797.
Grupp, H. (1989) *The Measurement of Technical Performance in the Framework of R&D Intensity, Patent and Trade Indicator*, FhG, Karlsruhe, mimeo.
Grupp, H. (1990) The concept of entropy in scientometrics and innovation research, *Scientometrics*, vol. 18, n. 3–4, pp. 219–239.
Grupp, H. and H. Legler (1989) *Strukturelle und Technologische Position der Bundersrepublik Deutschland im Internationalem Wettbewerb*, 1986–87, FhG, Hannover.
Grupp, H. and B. Schwitalla (1989) *Technometrics, Bibliometrics, Econometrics and Patent Analysis: Towards a Correlated System of Science, Technology, and Innovation Indicators*, in Van Raan et al., 1989.
Guerrieri, P. and C. Milana (1990) *L'Italia e il Commercio Mondiale*, Il Mulino, Bologna.
Hagedoorn, J. and J. Schakenraad (1992a) Leading companies and networks of strategic alliances in information technologies, *Research Policy*, vol. 21, n. 2, pp. 163–190.
Hagedoorn, J. and J. Schakenraad (1992b) *Strategic Technology Partnering and International Corporate Strategies*, forthcoming in Hughes, 1992.
Holland, S. (1987) *The Global Economy*, Weidenfeld and Nicolson, London.
Hood, N. and J.E. Vahlne (eds.) (1987) *Strategies in Global Competition*, Croom Helm, London.
Hughes, K. (1986) *Technology and Exports*, Cambridge University Press, Cambridge.
Hughes, K. (ed.) (1992) *European Competitiveness*, Cambridge University Press, Cambridge, forthcoming.
Irvine, J. and B.R. Martin (1989) International comparisons of scientific performance revisited, *Scientometrics*, vol. 15, n. 5–6, pp. 369–392.
Kaldor, M. (1981) *The Baroque Arsenal*, André Deutsch, London.
Kline, G.J. and N. Rosenberg (1986) *An Overview of Innovation*, in Landau-Rosenberg, 1986.

Kristensen, P.H. and J. Levinsen (1983) *The Small Countries Squeeze*, Forlaget for Samfundsokonomi og Planlaegning, Roskilde.
Krugman, P. (1979) A model of innovation, technology transfer and the world distribution of income, *The Journal of Political Economy*, vol. 87, n. 2, pp. 253–266.
Landau, N. and N. Rosenberg (eds.) (1986) *The Positive Sum Strategy. Harnessing Technology for Economic Growth*, National Academy Press, Washington D.C.
Levin, R.C. et al. (1984) *Survey Research on R&D Appropriability and Technological Opportunity. Part I – Appropriability*, Yale University, New Haven.
Levin, R.C. et al. (1987) Appropriating the returns from industrial research and development, *Brookings Papers on Economic Activity*, n. 3, pp. 783–831.
Leydersdorff, L. (1988) Problems with the "measurement" of national scientific perfomance, *Science and Public Policy*, vol. 15, n. 3, pp. 149–152.
Lindsey, D. (1989) Using citation counts as a measure of quality in science. Measuring what's measurable rather than what's valid, *Scientometrics*, vol. 15, n. 3–4, pp. 189–203.
Lundvall, B.A. (1985) Product innovation and user-producer interaction, *Industrial Development Research Series*, n. 31, Aalborg University Centre, Aalborg.
Luukkonen, T., O. Persson and G. Sivertsen (1992) An outline for understanding patterns of international scientific collaboration, *Science, Technology & Human Values*, vol. 17, n. 1, pp. 101–126.
Maddison, A. (1987) Growth and slowdown in advanced capitalist economies, *Journal of Economic Literature*, vol. 25, pp. 649–698.
Malaman, R. (ed.) (1991) *Brevetto e Politica dell'Innovazione*, Il Mulino, Bologna.
Mansfield, E. et al. (1982) *Technology Transfer, Productivity and Economic Policy*, Norton, New York.
Martin, B.R., J. Irvine, F. Narin and C. Sterritt (1987) The continuing decline of British science, *Nature*, vol. 330, 12 November.
Martin, B.R., J. Irvine and P. Isard (1991) *International Trends in Government Funding of Academic and Related Research*, SPRU, Brighton.
Melman, S. (1983) *Profits Without Production*, Knopf, New York.
Miquel, J.F. et al. (1989) Les scientifiques sont-ils ouverts à la coopération internationale?, *La Recherche*, n. 206, January.
Napolitano, G. and G. Sirilli (1990) The patent system and the exploitation of inventions: results of a statistical survey conducted in Italy, *Technovation*, vol. 10, n. 1, pp. 5–16.
Narin, F. and E. Noma (1985) Is technology becoming science?, *Scientometrics*, vol. 7, n. 3–6, pp. 369–381.
Narin, F. and D. Olivastro (1987a) *Identifying Areas of Strength and Excellence in U.K. Technology*, Department of Trade and Industry, London.
Narin, F. and D. Olivastro (1987b) *Identifying Areas of Strength and Excellence in F.R.G. Technology*, Bundesministerium für Forschung und Technologie, Bonn.
Narin, F. and D. Olivastro (1988a) *Identifying Areas of Leading Edge Japanese Science and Technology*, National Science Foundation, Washington, D.C.
Narin, F. and D. Olivastro (1988b) *Technology Indicators Based on Patents and Patent Citations*, in Van Raan, 1988.
Narin, F. and E.S. Whitlow (1990) *Measurement of Scientific Cooperation and Coauthorship in CEC-Related Areas of Science*, Commission of the European Communities, Brussels.
National Science Board (1989) *Science and Engineering Indicators*, USGPO, Washington D.C.
Nelson, R. (1981) Research on productivity growth and productivity differences: dead ends and new departures, *Journal of Economic Literature*, vol. 19, n. 3, pp. 1029–1064.
Nelson, R. (ed.) (1982) *Government and Technical Progress. A Cross-Industry Analysis*, Pergamon Press, New York.
Nelson, R. (1984) *High Technology Policies: A Five-Nations Comparison*, The American Enterprise Institute, Washington D.C.
Nelson, R. (1989) U.S. technological leadership: where did it come from and where did it go?, *Research Policy*, vol. 19, pp. 117–132.
Nelson, R. (ed.) (forthcoming) *National Innovation Systems: A Comparative Study*, Columbia Uni-

versity, mimeo.
Nelson, R. and S. Winter (1977) In search of a useful theory of innovation, *Research Policy*, vol. 5, n. 6, pp. 36–76.
Nelson, R. and S. Winter (1982) *An Evolutionary Theory of Economic Change*, Belknap Press of Harvard University Press, Cambridge, Mass.
OECD (1985a) *Science and Technology Policy Outlook*, Paris.
OECD (1990) *Report of the Technology and Economy Program*, Paris.
OECD (various years) *Main Science and Technology Indicators*, Paris.
Paci, R. (1990) *Specializzazione Tecnologica e Competitivita' Internazionale dell'Industria Italiana*, "Quaderni dell'Istituto di Scienze Economiche e Finanziarie", n. 1, Universita' di Cagliari, Cagliari.
Patel, P. and K. Pavitt (1987a) *Is Western Europe Losing the Technological Race?*, in Freeman, 1987b.
Patel, P. and K. Pavitt (1987b) *The Technological Activities of the U.K.: A Fresh Look*, British Association for the Advancement of Science, Belfast.
Patel, P. and K. Pavitt (1989a) *Do Large Firms Control the World's Technology?*, SPRU, Brighton.
Patel, P. and K. Pavitt (1989b) A comparison of technological activities in West Germany and the United Kingdom, *National Westminster Bank Quarterly Review*, May, pp. 27–42.
Patel, P. and K. Pavitt (1990) The importance of the technological activities of the world's largest firms, *World Patent Information*, vol. 12, n. 2, pp. 89–94.
Patel, P. and L. Soete (1988) *International Comparisons of Activity in Fast-Growing Patent Fields*, SPRU, Brighton.
Pavitt, K. (1984) Sectoral patterns of technical change: towards a taxonomy and a theory, *Research Policy*, vol. 13, n. 6, pp. 343–373.
Pavitt, K. (1987) *International Patterns of Technological Accumulation*, in Hood-Vahlne, 1987.
Pavitt, K. (1988) *Uses and Abuses of Patent Statistics*, in Van Raan, 1988.
Pavitt, K. and P. Patel (1990) Sources and direction of technological accumulation in France: a statistical comparison with Germany and the UK, *Technology Analysis and Strategic Management*, vol. 2, n. 9, pp. 3–26.
Pavitt, K. and L. Soete (1981) International differences in economic growth and the international location of innovation, in Giersh, H., *Emerging Technologies*, Tübingen.
Petrella, R. (1989) Globalization of technological innovation, *Technology Analysis and Strategic Management*, vol. 1, n. 4, pp. 393–407.
Pianta, M. (1988a) High technology programmes: for the military or for the economy?, *Bulletin of Peace Proposals*, vol. 19, n. 1.
Pianta, M. (1988b) *New Technologies across the Atlantic: US Leadership or European Autonomy?* Harvester-Wheatsheaf, The United Nations University, Hemel Hempstead, Tokyo.
Pianta, M. (1989) *Internazionalizzazione della Tecnologia e Processi Innovativi nei Paesi piu' Avanzati*, Rapporto ENEA, Rome.
Pianta, M. and D. Archibugi (1991) La specializzazione tecnologica italiana nel contesto internazionale: un'analisi dei brevetti, *L'Industria*, vol. 12, n. 3, pp. 397–428.
Pianta, M. and R. Simonetti (1990) *La Specializzazione Scientifica dei Paesi piu' Avanzati: Un'Analisi degli Indicatori Bibliometrici*, CNR-ISRDS, Technical Report, Rome.
Piore, M.J. and C. Sabel (1984) *The Second Industrial Divide*, Basic Books, New York.
Porter, M. (1990) *The Competitive Advantage of Nations*, MacMillan, New York.
Posner, M. (1961) International trade and technical change, *Oxford Economic Papers*, vol. 13, n. 3, pp. 323–341.
Price, D.J. De Solla (1984) The science/technology relationship, the craft of experimental science, and policy for the improvement of high technology innovations, *Research Policy*, vol. 13, n.1 pp. 3–20.
Price, D.J. De Solla (1986) *Little Science, Big Science ... and beyond*, Columbia University Press, New York.
Reich, R. (1987) The rise of techno-nationalism, *The Atlantic Monthly*, May.
Roobeek, J.M. (1990) *Beyond the Technology Race*, Elsevier, Amsterdam.
Rosenberg, N. (1976) *Perspectives on Technology*, Cambridge University Press, Cambridge.
Rosenberg, N. (1982) *Inside the Black Box: Technology and Economics*, Cambridge University Press,

Cambridge.
Rosenberg, N. (1986) *Civilian 'Spillovers' from Military R&D Spending: The American Experience since World War II*, Stanford University, September 1986.
Sassu, A. and R. Paci (1989) Brevetti di invenzione e cambiamento tecnologico in Italia, *Rivista di Politica Economica*, vol. 89, n. 1, pp. 73–123.
Savignon, F. et al. (1982) *Les Dépôts à l'étranger: Un Indicateur de Qualité des Brevets?*, OECD, Paris.
Schankerman, M. and A. Pakes (1986) Estimates of the value of patent rights in European countries during the post-1950 period, *Economic Journal*, vol. 96, December, pp. 1052–1076.
Scherer, F.M. (1983) The propensity to patent, *International Journal of Industrial Organization*, vol. 1, pp. 107–128.
Scherer, F.M. (1984) *Innovation and Growth. Schumpeterian Perspectives*, MIT Press, Cambridge, Mass.
Schiffel, D. and C. Kitti (1978) Rates of invention: international patent comparisons, *Research Policy*, vol. 7, n. 4, pp. 324–340.
Schmoch, U. et al. (1988) *Technikprognosen mit Patentindikatoren*, Verlag TUV Rheinland GmbH, Köln.
Schmoch, U. and H. Grupp (1989) *Patent between Corporate Strategy and Technology Output: An Approach to the Synoptic Evaluation of US, European and West German Patent Data*, in Van Raan et al., 1989.
Servan-Schreiber, J.J. (1968) *The American Challenge*, Penguin, Harmondsworth.
Simonetti, R. (1992) Un'analisi per settori della specializzazione tecnologica: i dati sui brevetti dei paesi avanzati, *Economia e Politica Industriale*, forthcoming.
Sirilli, G. (1987) *Patents and Inventors: An Empirical Study*, in Freeman, 1987b.
Small, H. and B.C. Griffith (1974) The structure of scientific literature I: identifying and graphing specialities, *Social Studies of Science*, vol. 4, n. 1, pp. 17–40.
Soete, L. (1981) A general test of the technology gap trade theory, *Weltwirschaftliches Archiv*, vol. 117, pp. 638–660.
Soete, L. (1987) *The Impact of Technological Innovation on International Trade Patterns: The Evidence Reconsidered*, in Freeman, 1987b.
Soete, L. (1988) *Technical Change and International Implications for Small Countries*, in Freeman-Lundvall, 1988.
Soete, L. and S. Wyatt (1983) The use of foreign patenting as an internationally comparable science and technology indicator, *Scientometrics*, vol. 5, pp. 31–54.
Stokes, T.D. and J.A. Hartley (1989) Coauthorship, social structure and influence within specialities, *Social Studies of Science*, vol. 19, pp. 101–125.
Thurow, L. (1985) *The Zero-Sum Solution: Building a World-Class American Economy*, Simon and Schuster, New York.
Tirman, J. (ed.) (1984) *The Militarization of High Technology*, Ballinger, Cambridge.
Trajtenberg, M. (1990) *Economic Analysis of Product Innovation. The Case of CT Scanners*, Harvard University Press, Cambridge, Mass.
US National Science Board (1989) *Science and Engineering Indicators 1989*. USGPO, Washington, D.C.
US Patent and Trademark Office (1985, 1989) *Concordance: United States Patent Classification to International Patent Classification*, US Government Printing Office, Washington D.C.
US Patent and Trademark Office (1985) *Review and Assessment of the OTAF Concordance between the U.S. Patent Classification and the Standard Classification Systems*, USGPO, Washington, D.C.
US Office of Technology Assessment (1988) *Technology and the American Economic Transition: Choices for the Future*. OTA-TET-283, USGPO, Washington D.C.
Van Raan, A.F.J. (ed.) (1988) *Handbook of Quantitative Studies of Science and Technology*, Elsevier, Amsterdam.
Van Raan, A.F.J., A.J. Nederhof and H.F. Moed (eds.) (1989) *Science and Technology Indicators: Their Use in Science Policy and Their Role in Science Studies*, DSWO Press, Leiden.
Van Raan, A.F.J. and R.J.W. Tijssen (1990) *An Overview of Quantitative Science and Technology*

Indicators Based on Bibliometric Methods, Paper to the OECD Conference, Paris, 2–5 July.
Van Tulder, R. and G. Junne (1988) *European Multinationals in Core Technologies*, Wiley, London.
Van Vianen, B.G., H.F. Moed and A.F.J. Van Raan (1990) An explanation of the science base of recent technology, *Research Policy*, vol. 19, n. 1, pp. 61–81.
Vernon, R. (1981) *Technology Effects on International Trade: A Look Ahead*, in Giersh, 1981.
Von Hippel, E. (1988) *The Sources of Innovation*, Oxford University Press, New York.
Von Tunzelmann, N. (1989) *Convergence of Firms in Information and Communication: A Test Using Patents Data*, SPRU, Brighton.
Walsh, V. (1984) Invention and innovation in the chemical industry: demand pull or discovery push?, *Research Policy*, vol. 13, n. 4, pp. 211–234.
Walsh, V. (1988) *Technology and the Competitiveness of Small Countries: A Review*, in Freeman-Lundvall, 1988.
Weil, V. and W.J. Snapper (eds.) (1989) *Owning Scientific and Technical Information*, Rutgers University Press, New Brunswick and London.
Wheal, P.R. and R.M. McNally (1986) Patent trend analysis. The case of microgenetic engineering, *Futures*, October.
Winter, S. (1989) *Patents in Complex Contexts: Incentives and Effectiveness*, in Weil-Snapper, 1989.
Wyatt, S., G. Bertin and K. Pavitt (1985) Patents and multinational corporations: results from questionnaires, *World Patent Information*, vol. 7, n. 3, pp. 196–212.

Index

SPRINGER NATURE

GPSR Compliance

The European Union's (EU) General Product Safety Regulation (GPSR) is a set of rules that requires consumer products to be safe and our obligations to ensure this.

If you have any concerns about our products, you can contact us on ProductSafety@springernature.com

In case Publisher is established outside the EU, the EU authorized representative is:

Springer Nature Customer Service Center GmbH
Europaplatz 3
69115 Heidelberg, Germany

FSC
www.fsc.org
MIX
Papier | Fördert
gute Waldnutzung
FSC® C083411

Kolibri 360 GmbH
Ferdinand-Jühlke-Straße 7
99095 Erfurt, Deutschland
produktsicherheit@kolibri360.de